ROYAL AGRICULTURE

Scrapie Disease in Sheep

Scrapie Disease in Sheep

Historical, Clinical, Epidemiological, Pathological and Practical Aspects of the Natural Disease

H. B. PARRY

Nuffield Institute for Medical Research,
University of Oxford, Oxford, UK

Edited by

D. R. OPPENHEIMER

Department of Neuropathology,
The Radcliffe Infirmary, Oxford, UK

1983

Academic Press

A Subsidiary of Harcourt Brace Jovanovich, Publishers

London   New York
Paris   San Diego   San Francisco   São Paulo
Sydney   Tokyo   Toronto

ACADEMIC PRESS INC. (LONDON) LTD.
24–28 Oval Road
London NW1 7DX

U.S. Edition published by
ACADEMIC PRESS INC.
111 Fifth Avenue
New York, New York 10003

British Library Cataloguing in Publication Data
Parry, H. B.
Scrapie disease in sheep.
1. Sheep—Diseases   2. Scrapie
I. Title
636.3′0896925   SF968

ISBN 0-12-545750-2
LCCCN 83-71359

Typeset by Bath Typesetting Ltd., Bath
Printed in Great Britain by Page Bros. (Norwich) Ltd.

Sheep of Suffolk breed, with advanced scrapie. The lower picture shows pigmentation of denuded areas.

Preface

At the time of his sudden death in 1980, H. B. (James) Parry was engaged in writing a massive monograph on the subject which had occupied most of his time and energy for the past 25 years—that is, Natural Scrapie of Sheep. The projected work was designed to occupy some 1200 pages of typescript (about 750,000 words) excluding the bibliography and various appendices. Of this, barely a quarter was already written and typed. For the rest, Parry's intentions were clear from his projected chapter headings and sub-headings.

Looking at the unfinished text, a group of Parry's friends and associates agreed that the material contained in it was far too valuable to be lost. We had several reasons for thinking this, other than the obligations of friendship. The first was that the work contained a detailed account of the history of the disease in Europe and elsewhere, over the past 250 years. This was the outcome of several years of scholarly enquiry in the archives of various European centres of learning, and the amassing of a unique collection of letters and of photocopies taken from obscure 18th and 19th century publications. The result of this study is not merely a work of scholarship; it is also an important piece of epidemiological research, which will have to be taken into account by those responsible for future policy-making in the control of scrapie, whatever view may be taken on the nature and causation of the disease.

Our second reason was that although a great deal has now been written about the laboratory investigation of artificial scrapie, there is relatively little on the clinical and epidemiological aspects of the naturally occurring disease. Parry, as a working veterinarian, had accumulated an unrivalled experience of scrapie in the field. For many years there has been and there still remains a stigma about this disease which makes flockmasters and shepherds reluctant to discuss the subject with outsiders, including apparently disinterested research workers. Parry was able to overcome this reluctance, and succeeded in establishing over the years a close, almost symbiotic, relationship with a number of important flock-owners. Under Parry's direction, accurate and detailed pedigrees and flock-breeding records were kept, and the occurrence of scrapie accurately noted, the affected animals being taken over by Parry for test-mating and post-mortem examination. Parry provided free consultant service on day-to-day health problems; more important, from the owners' point of view, was the fact that following Parry's genetic counselling, scrapie practically disappeared from their flocks.

The third reason why we thought it was important to publish this work is somewhat more difficult to explain. The 'official' view—that is, the view which is held in the important publicly-funded research centres—is that scrapie is a contagious disease, due to the action of a so-called 'slow virus'. This view is based mainly on the fact that the injection of tissue extracts from affected animals into other sheep, or into a variety of laboratory animals, results, after a certain delay, in the appearance of a somewhat similar disease in the recipient; tissue extracts from this animal can produce similar disease in a third animal, and so on—in other words, there is, associated with scrapie, a self-replicating transmissible agent. From this piece of evidence, it is natural—some would say inevitable—to infer that the disease is acquired (as tuberculosis or foot-and-mouth disease is acquired) by infection. Parry, while not disputing the existence of the transmissible agent, was nevertheless convinced, from his experience in the field, that under natural conditions the disease was transmitted not by an infectious agent but by a Mendelian recessive gene. The idea that a disease could be at the same time hereditary and potentially infective was novel and disturbing, and seemed to involve the creation of a completely new category of disease. In view of our vast ignorance (a) of the physicochemical structure of the scrapie transmissible agent and (b) of the mechanisms by which aberrant genes exert their damaging effects, it should not have been too difficult for both sides to keep an open mind on the theoretical problem, or even to devise a plan of co-operative research to settle the matter. For various reasons, this did not happen; what is worse, a bitter hostility arose between Parry and the 'official' group. Even if I knew more of the details of this unfortunate dispute, I should not regard it as my business to apportion the blame. What is important, and lamentable, is that, as a result, the work of Parry and of his associates tended to be ignored in the official publications. Now that Parry is dead, it is to be hoped that animosities will subside, and that serious attention will be paid to his work, with its merits and its shortcomings. Meanwhile, the publication of this truncated monograph will do something, I hope, to redress a certain imbalance in the current literature.

It remains to explain my role in editing this book. Faced with the uncompleted work, I thought at first that the only course was to publish the completed chapters as separate essays. I soon realised that this would not do justice to Parry's intentions, and that a coherent, continuous text could be achieved, almost entirely by scissors-and-paste methods. This was largely because the original chapter 2, which was complete, contained a synopsis of the subsequent chapters, including the final chapter, which was to have contained Parry's conclusions and recommendations. The chapter was accordingly dismembered: parts were grafted onto other chapters, and parts were discarded as repetitions; the remainder forms the bulk of the final chapter of the book as it stands. The original chapter 3, dealing with the history of scrapie in Europe and elsewhere, has been cut in two, and the material rearranged. The original chapters 6, 7, 8 and 9 have become chapters 4, 5, 6 and 7, with minor alterations. The original chapters 21 and 22 are incorporated into chapter 8. The remaining chapters (the original plan was for a total of 29 chapters) cannot now be reconstructed. It is clear from the headings and notes that they were to be largely speculative, concerned with broad biological issues rather than with detailed discussion of natural scrapie.

I have to confess to a certain amount of tampering with Parry's text. In general, I have done this in the interests of clarity and readability. I do not think I have anywhere altered the meaning of what he wrote; if I have sinned in this way, I apologise to the reader and to Parry's memory. The original text has not been destroyed, and can still be consulted, along with Parry's other papers, at the library of the Institute of Agricultural History in Reading.

Finally, a word is needed to explain what some readers might regard as a wilful neglect of recent literature on scrapie—in particular, the large volume of published work on the experimental disease. The explanation is very simple: Parry had planned to write on these matters, but died before doing so.

I gratefully acknowledge the help I have received from various colleagues and friends; in particular Dr. Gerald Draper, Dr. David Smith and Dr. Derek Wyatt. I am also grateful to James's widow,

Honor Parry, for her help and encouragement; to the Sheep Development Association, who have generously defrayed the secretarial expenses; to Miss Barbara Newton for her help with the bibliography; to Mrs. Elizabeth Beck for her help with the micrographs; and to Mr. Ken Brown for his drawings of sheep.

Trinity College, Oxford   D. R. Oppenheimer
June, 1983

Foreword

When I learned of James Parry's death my grief for the loss of a friend was intensified by a concern that his major scientific undertaking, to which he had devoted so much of his working life, would now never be published. His work had largely been incorporated into the manuscript of a book on scrapie which had been slowly accumulating over many years. That it has now been made ready for publication we owe to Dr. David Oppenheimer. He is to be congratulated on his sensitive editing of the manuscript. Though some sections of the text have had to be pruned, I believe that none of the essential features of James Parry's original work have been omitted. In particular, the confidential data gathered so painstakingly from sheep flocks and their owners over many years have been preserved. Though these data are important to science, and that is the main reason now for their publication, this was not the only reason for which they were originally collected. James Parry was able to make predictions on the basis of his findings which were of great help to his colleagues, the sheep breeders, and, moreover, secured their continual cooperation in his project.

The scientific polemics that surrounded James Parry's work were unfortunate and, one would have thought, unnecessary. It is an acknowledged part of scientific method to present a clear viewpoint or hypothesis from which conclusions can be drawn and predictions made. In throwing forth a position, or hypothesis, one automatically draws attention to complementary or opposing ideas or hypotheses; and one perforce allows that an opposing idea has a right to be heard and a chance of being right; moreover, unless one is dealing with logical incompatibilities, one must accept the possibility of truth being a synthesis. In reference to scrapie one might have innocently thought that these considerations would have been even more applicable since questions had been raised about the properties of the scrapie agent which shook the most basic biological concepts. Yet the petty squabbles continued. Perhaps in the discourse between scientific colleagues, as in other human relationships, there is a fixed sum of logic and emotion.

I think we must accept the fact that there is no clear dividing line between genes and viruses. If this is allowed, it should come as no surprise that a genetic disease may prove to be transmissible under experimental conditions or occasionally transmitted horizontally under natural conditions; nor that a virus (or its transcribed equivalent) which causes a respectable infection by horizontal transmission may end up in a host's genome. This of course assumes that one does not insist on thinking of a 'virus' purely in morphological terms; or of a 'gene' as part of the genome which must be long enough to code for a protein. In the field of scrapie one would have thought that these were not likely sources of misapprehension. The properties, nature, indeed the very existence, of the scrapie agent were constantly being questioned. Yet controversy continued to flare up around James Parry: vertical or horizontal transmission; gene or virus. Doubt it seems is as productive of dogmatic as of sceptical responses.

For some of us it was easy to see how both views might be true. There are a number of ways of explaining and reconciling the data; they are mostly speculative, yet scrapie and the other subacute spongiform virus encephalopathies (transmissible mink encephalopathy, Creutzfeldt-Jakob disease and kuru) demand speculation. The virus may be transmitted genomically in a repressed form with only a very low probability of expression. Such a gene may in fact be ubiquitous—but this does not mean that its expression must be universal, and the failure to find scrapie infectivity in normal animals does not imply that the scrapie gene is not carried by them. Repressed genomic transmission might explain, for example, the rare sporadic event throughout human populations of a person coming down with the human equivalent of scrapie, Creutzfeldt-Jakob disease. The genetically determined disease might be expressed through a translated protein or polypeptide; but then again it may depend on release or transcription from the genome of nucleic acid which, suitably packaged in host protein, might spread from cell to cell to produce slowly progressive degenerative disease. Such an agent would be similarly infectious if, by chance or experimental design or inadvertently through host behaviour, it became transmitted horizontally to another host. In the context of scrapie one must remember that it has been further speculated that the agent is the protein product of the derepressed gene and that it is this protein, hormone-like, with no genetic material, which spreads horizontally from cell to cell causing disease and may transmit disease to another host.

Since the subacute spongiform virus encephalopathies are ubiquitous and yet rare, except under special circumstances (such as the epidemic of kuru in a population practising endocannibalism), the event or events leading to the development of disease and the spread from cell to cell of the infectious agent must be improbable; and therefore unlikely mechanisms become most permissible explanations. Thus one has to allow the possibility that scrapie may very, very rarely spread from sheep to man to produce Creutzfeldt-Jakob disease, even though the distribution of scrapie in the world makes it clear that this cannot be the normal means of transmission of Creutzfeldt-Jakob disease. The genetic interaction between the scrapie agent and its hosts implies nucleic acid in its core. This may well have escaped from a genome in the production of the natural disease. In certain familial clusters of Creutzfeldt-Jakob disease and in certain breeds of sheep it would seem that the gene is transmitted with high expressivity, allowing for Mendelian patterns of inheritance to be observed. Mendelian recessiveness may, of course, be explained by the expression of a universal gene only in the complete absence of controlling genes. Scrapie may also occur sporadically in other breeds of sheep and if it were as rare in sheep as Creutzfeldt-Jakob disease in its sporadic form is in man then it would not be detected. Thus the patterns observed by James Parry in breeding flocks of high scrapie incidence, for which he produces strong evidence, are by no means irreconcilable with the patterns of disease observed in experimental scrapie or in Creutzfeldt-Jakob disease. The variation found in scrapie is indeed mirrored by the wide clinical variation of Creutzfeldt-Jakob disease. Yet where an epidemic occurs the disease is remarkably invariant. Examples are transmissible mink encephalopathy, which is said to have originated from scrapie-infected meat fed to mink, and kuru, which arose, we think, from a sporadic case of the ataxic form of Creutzfeldt-Jakob disease occurring in the New Guinea highlands (the double unlikelihood of this happening in a group practising endocannibalism is of no relevance once the event has occurred, when the sequence of events becomes devastatingly predictable). Kuru is stereotyped in its clinical and pathological presentation, unlike the range found in Creutzfeldt-Jakob disease, and this is consistent with kuru being a particular genetic strain of Creutzfeldt-Jakob disease which breeds true.

The school which considers the scrapie agent to be a hormone (in a manner of speaking) is by no means dead, and this is consistent with the view of the matter taken by James Parry himself. Some of these ideas are difficult to reconcile with standard biological concepts and certainly with the standard properties of virus (which of course constitute a descriptive artifact, however entrenched they may be in medical or scientific minds). But these ideas are not inherently ridiculous, and the most exciting thing about this field of unconventional slow virus infections over many years is that it has compelled us to keep an open mind on all these issues. Although this may be generally true, unfortunately it was not true of everyone working in the field. We must be grateful to James Parry for keeping open an important aspect of the scrapie story and working hard to produce rigorous data to illuminate it. The publication of these results is a tribute to his memory as a man and his assiduity as a scientist.

Institute of Medical Research,   **Michael Alpers**
Goroka, Papua New Guinea.
January, 1983.

Contents

Herbert Butler (James) Parry

1912	Born in Calne, Wiltshire, of farming stock.
1930–33	Queens' College, Cambridge. Natural Sciences Tripos. First class honours in Part 2 Physiology. B.A. 1933; M.A. 1948.
1935–38	Edinburgh and London Veterinary Colleges.
1938	Member of Royal College of Veterinary Surgeons (M.R.C.V.S.)
1938–40	Commonwealth Fund Fellowship in Biochemistry and Nutrition at University of Wisconsin, USA.
1940–43	Lecturer and Senior Lecturer in Veterinary Medicine, University of Sydney.
1943–47	Research Staff, Imperial Chemical Industries Limited, Dyestuffs Division, Manchester.
1947–53	Senior Scientific Officer, Animal Health Trust, Canine Research Station, Newmarket, Cambridgeshire. Investigated night blindness in Irish Red Setters.
1953–80	Senior Research Worker, Nuffield Institute for Medical Research, Oxford, studying diseases of sheep.
1953–56	In collaboration with Heather Shelley, demonstrated the dietary causes of pregnancy toxaemia in sheep. Research concentrated upon scrapie from this time on.
1965	Original Fellow of Iffley (later Wolfson) College, Oxford. Active member of the Building Committee of the new Wolfson College.
1967	George Hedley Memorial Award for outstanding service to the sheep industry.
1973	Organised Wolfson College Lectures on problems of world population (book: *Population and its Problems; a plain man's guide*, ed. H. B. Parry, Oxford, 1974).
1980	Died from a heart attack.

Subjects of published articles from 1948 onward include:
canine hysteria, epilepsy in dogs, virus hepatitis of dogs, progressive retinal atrophy in dogs, haemolytic disease of foals, and toxaemia of pregnancy in sheep. For published articles on scrapie, written alone or in collaboration, see items 22, 23, 37, 38, 39, 40, 43, 107, 184, 185, 224, 225, 226, 227, 228, 229, 230, 231 and 232 in the bibliography at the end of this book.

Chapter 1

General Conspectus

I. The Scrapie Group of Diseases

Scrapie, as an untreatable slowly fatal disease of sheep, has been known in many parts of Europe for 250 years. Its cause, cure and prevention have been and remain a matter of controversy between those who favour an infectious cause and those who emphasise the importance of heredity. Since the demonstration of an experimentally transmissible agent in the tissues of an affected sheep in 1936 the general view has favoured an infectious aetiology. Since 1950 much information has become available from laboratory studies of scrapie and from field experience, as well as from human medicine, which makes a new evaluation of all our present knowledge appropriate.

This is particularly so since scrapie is now generally accepted as the representative type disease for a group of neurological disorders of man and of domestic animals, characterised by:

(1) insidious onset, without prior ill-health, of a progressive, fatal neurological illness of somewhat variable expression, appearing most commonly in middle age and later;

(2) the development of a characteristic non-inflammatory degeneration, bilaterally symmetrical, affecting specific portions of the nervous system, with disappearance (or 'fall-out') of nerve cells, which may exhibit cytoplasmic vacuolation, and microcystic (spongiform) changes in the neuropil, associated with glial proliferation;

(3) the presence in tissue homogenates, but not in blood, secretions or excreta, of an experimentally transmissible agent or factor capable of inducing on injection a somewhat similar transmissible neurocytopathic disease in laboratory animals; and

(4) a pattern of occurrence which is frequently familial, sometimes suggesting an hereditary predisposition, though compelling evidence for a hereditary factor is usually unavailable or unsought.

The transmissible agent, or agents, all of which produce a remarkably similar neurocytopathic reaction, are often called 'slow viruses', but remain ill-defined, since (a) no immunological characteristics, either humoral or cellular, have been demonstrated; (b) electron microscopic evidence for a virus-like morphological structure is lacking; (c) natural communicability in the field has not been established, with a possible exception in the case of kuru in Papua New Guinea, which is thought to have been transmitted by ritual cannibalism; (d) their presence is only demonstrable by animal inoculation of tissue fragments, which is followed after a protracted latent period by the development of spongiform encephalopathic changes.

The diseases usually considered to fall within this group are (1) in animals, scrapie of sheep, the rarer form of scrapie in goats, and the transmissible encephalopathy of mink and (2) in man, kuru of the Foré and related peoples of the Central Highlands of New Guinea, and a group of spongiform encephalopathies, of worldwide distribution, commonly included under the heading of Creutzfeldt-Jakob disease.

In general, the kuru of Papua New Guinea excepted, the causes and modes of dissemination of these diseases remain uncertain, and specific treatment and methods of prevention are unknown. Studies of their detailed natural history present great difficulties. The biological importance of scrapie lies in the contribution which studies of the epidemiological and biological features of the disorder in sheep may make to the general elucidation of these degenerative neurological disorders in which both hereditary and experimentally transmissible factors may be involved.

II. The Nature and Manifestation of Scrapie in Sheep

Scrapie manifests itself as a disorder of behaviour, locomotion and body homoeostasis. It occurs as an age-related disorder, normally first appearing between the ages of 2 and 5 years, progressing slowly, over 3 to 6 months or longer, to a fatal outcome.

Although the catalogue of signs or disabilities associated with the disease appears to be complex, in practice the disease presents with clearly recognisable patterns of signs (i.e. syndromes), based on combinations of five main functional disturbances: (1) *inanition* or loss of body weight, and especially of skeletal muscle mass, but with occasional obesity; (2) *ataxia* or clumsiness of bodily movements; (3) compulsive *rubbing* or *nibbling* of the dorsal lumbar region of the tail-base, the lateral thorax, or poll of the head, and nibbling of the haired portions of the limbs; (4) *mental changes* affecting behaviour and emotional responses; and (5) loss of fine control of *body homoeostasis*.

The development and progression of these disturbances are essentially gradual and continuous, although frequently so slight as to be unremarked for several months in a case with a slow rate of progress. The usual order of appearance is as follows:

(i) The earliest signs are often behavioural, with slight apprehensiveness, general restlessness, distrust of man and a failure to respond to his dog, a staring expression, and an elevated head posture referred to in the older literature as 'shrugginess'. The ears are often floppy, and the wool tip whitish. Such inconspicuous signs frequently go unnoticed unless the observer is experienced.

(ii) Occasional rubbing, usually of the base of the tail, the lateral thorax, or the poll of the head, and later compulsive nibbling at the lower portions of the legs.

(iii) Clumsiness of movements, especially of the hind quarters when turning, which proceeds to serious inco-ordination, a 'trotting' gait and a tendency to fall over if hustled, with signs of a posterior limb palsy without loss of tendon reflexes.

(iv) A fine trembling of the body and/or head, associated with inability to maintain body posture without substantial coarsely adjusted movements of the postural muscles.

(v) A loss of body weight and of skeletal muscle mass, i.e. wasting, with weakness, but without loss of appetite or obviously diminished food intake. Occasionally animals become excessively large, probably owing to increased skeletal muscle mass as well as to obesity. There is progressive lowering of exercise tolerance, with early fatigability on being driven. Frequently water and salt intakes are abnormal, and drinking habits are distorted.

(vi) Other signs, such as blindness, epileptiform convulsions, sudden

posterior palsy, may occur, and those listed above may be so inconspicuous as to go undetected. The disorder thus occurs as a variable
constellation of signs; most cases, however, exhibit the triad of signs of
rubbing, ataxia and wasting.
(vii) Careful observers have from 1750 onwards distinguished clearly
the scrapie (or the 'trotting disease', or 'rubbing disease') from parasitic
scab, sturdy or gid (coenurosis) and sporadic microbial meningo-
encephalitides, the three principal disorders with which it may be
confused.

The manifestation of clinical disease is age-related, usually appearing
between 2 and 5 years in most breeds and sheep populations; about
85% of all cases manifest by $4\frac{1}{2}$ years and 90% by 5 years. The male
manifests usually about 6 months younger than the female. Thus an
unaffected sheep has to be observed to $4\frac{1}{2}$ years old, and preferably
5 years or older, before it can be deemed to be unaffected, while still
retaining a 1 in 10 chance that it may become affected.

From the accounts available over 250 years the clinical manifestation
of the disease has changed but little. The small but characteristic
details recorded in the accounts of the 18th century can be immediately
recognised by an experienced observer of the mid-20th century. There
is no other syndrome in sheep with which scrapie, when fully developed,
is likely to be confused by knowledgeable observers. There is a remarkable unanimity in the accounts from Britain from 1750 onwards, and
especially in England between 1780 and 1810, from Germany in the
same period, and later from France and Scotland. The best extant
account of the disease remains that from Germany, assembled by
May (1868) (199) and given in extended English translation by
M'Gowan (1914) (204). Reading these accounts today, with recent
experience, one instantly recognises the complaint these older writers
were describing.

III. Ideas on Aetiology

From the earliest available account, that of the Memorial to the
British House of Commons in 1754–1755 (167), the cause of the malady
has been attributed to hereditary constitution by some and to an

infection by others. Each school of thought could point to much evidence compatible with its view; for heredity, its occurrence among the progeny of certain rams but not others (77) and with excessive inbreeding, in England (95), in Spain (289) and in Germany (199); for an infection the 'epidemics' which followed the introduction of certain animals, notably the introduction into France of Spanish merinos (129, 180, 189).

In the 1930's two important papers were published by French workers. In 1936 Cuillé and Chelle (87) set up, after a latent period of many months, a scrapie-like disease in a sheep by the inoculation of tissues from a sheep with natural scrapie. This observation was repeated in the goat (75), and confirmed in the sheep by many investigators (132, 323). In 1937 Bertrand, Carré and Lucam (33) published their careful study of the neuropathology of natural scrapie in sheep, and extended greatly our knowledge of the location, extent and nature (in addition to neuronal vacuolation) of the degenerative changes in the brain and spinal cord. They provided for the first time some pathological basis for the clinical neurological signs, although they did not draw attention to the distribution of the primary neuronal lesions in certain restricted neuro-anatomical systems reported in 1964 by Beck, Daniel and Parry (23). However, there were now firm enough histopathological criteria to support a diagnosis of natural scrapie.

The successful transfer of the disease by inoculation of tissue homogenates immediately stimulated an immense output of laboratory studies (see Kimberlin, 171), which sought to identify and characterise the nature of this transmissible but otherwise unidentifiable agent found in tissues. These studies were expedited greatly by transmission of the disease to mice (73), rats (74) and hamsters (335) and the reduction of the latent period before appearance of the disease from 12–15 months on first inoculation to 4–5 months, after some five transfers through laboratory animals. The artificially-induced disease in the mouse thus provided a cheap and convenient model of the disease, which was uncritically, although perhaps naturally, assumed to simulate closely in all important respects the natural disease in sheep. Indeed most of the literature on 'scrapie' refers to this artificial form of the disorder in small laboratory animals, and virtually all the theoretical constructions applied to 'scrapie' (128, 2, 159) are based on inferences derived from data acquired in small laboratory animals.

The object of the quest was the isolation and identification of the agent, generally considered at first to be an infectious virus of conventional type, against which it was hoped to develop a vaccine for use in immunisation against the natural disease in the field. Only later, when it became clear that the agent lacked the customary immunological characteristics of a conventional virus, was the agent termed a 'slow virus'. More recently, the term 'unconventional virus' has been used (124). Acceptable evidence of any virus-like morphological structure has not yet been found. Bignami and Parry (39) failed to find any electron microscopic evidence for the presence, in sheep with natural scrapie, of a morphologically distinct structure which might be identified as the transmissible agent, and subsequent observers have fared no better.

In spite of these failures to characterise the agent, the fact that the disease can be transmitted and serially passaged has been taken as proof that the natural disease is caused by the agent—in other words, that natural scrapie is an infectious disease, and should be treated as such. This, in fact, is the official view in most of the countries where scrapie presents a problem. The results of a few field studies (52, 100, 155) have been taken as evidence that animals from 'scrapie-free' flocks can be 'infected' by natural contact with scrapie-affected animals. These studies, however, are of doubtful significance in view of the uncertain meaning of the term 'scrapie-free'. Flockowners—in particular, breeders of pedigree stock—are naturally reticent, and reluctant to admit to the occurrence of scrapie among the animals which they handle. My own repeated attempts to demonstrate communicability of the disease by introducing sheep from affected flocks to unaffected flocks and vice versa have revealed no evidence to support the view that natural scrapie is ever transmitted in the field by an infectious agent through contact, coitus, suckling, contaminated surroundings or intermediary vectors (227, 228, 229).

My own view, based on 25 years' experience as a veterinary consultant to a number of large pedigree breeding establishments, and on a unique collection of detailed breeding records, is that infection plays little or no part in the production of the natural disease. On the contrary, the accumulated data from these observations are fully compatible with the view that natural scrapie is passed on by a simple autosomal recessive gene The practical outcome of this work has been the elimination of the disease, by selective breeding, from flocks which at the beginning of the period were suffering severe losses from scrapie (228). The details of the work are dealt with in later sections of this book.

It is to be hoped that an intellectually satisfying account of the aetiology of scrapie will be achieved during the next decade. Meanwhile, the dispute between upholders of the infective and the genetic theories continues. Much of the evidence which any satisfactory theory must embrace is to be found in the history of the disease, and its epidemiology—that is, in the study of when the disease has occurred, in what parts of the world, in what animal populations and in what conditions of sheep husbandry. The next two chapters contain the results of my enquiries into these matters. The information is still far from complete, but it is a great deal fuller than any available published account.

The fourth chapter describes the variety of clinical syndromes occurring in natural scrapie, and the fifth and sixth give an account of my own observations on the epidemiology of the disease. The seventh and eighth are concerned with pathology. Most of the information in these has already appeared in print, but is reproduced here in a fair amount of detail because it has been ignored in most of the currently available accounts of the pathology of scrapie. Chapter nine is concerned with a number of biological problems, including a discussion of the nature of so-called system degenerations in the central nervous system of man and other animals, and speculations on the role of neurophysins in scrapie and in other conditions. The final chapter contains my conclusions and practical recommendations for the control of scrapie.

Chapter 2

Historical Background

I. The Quality of the Data

The clinical forms and epidemiological characteristics of scrapie are such that the observations of perceptive owners and shepherds made 200 years ago are often immediately recognisable today by anyone who has had practical experience of the disease. Thus one is able to accept a diagnosis based on these clinical accounts, often illuminated by a characteristic aside on the practical aspects of the disease. This assessment is reinforced by the absence of any other disease of sheep recognised then or now, with which the disease is likely to have been confused by experienced and well-informed observers, who have recognised the distinction from coenurosis ('gid') and the encephalitis of louping ill from early times. Such observers were prepared to write of their experiences more fully in the 18th century than their successors have done in the present, when study of the disease has been left to professionally qualified outsiders, not in close touch with the sheep.

It is not mere chance that has ensured an extremely high standard in the accounts of the disease published in many parts of north-western Europe from 1750 onwards. The period was one of great intellectual questioning, of innovation and experimentation, often referred to as 'the Age of Enlightenment'. The innovatory ferment influenced agriculturists and landowners to a remarkable degree in many parts of Western Europe, and particularly affected livestock improvement and breeding. This response was facilitated, indeed made possible, by a number of supporting circumstances; the physical climate, after the 'little ice-age' of the 16th and 17th centuries, improved in north-western Europe (177), and was congenial to agriculture, in Britain at least, until 1879; the system of land tenure was such that individual minor gentry and yeoman farmers were in command of land holdings and financial resources commensurate with the special needs for genetic improvement of livestock; national economies and populations were expanding, and resources were available to provide what economists happily refer to as 'infra-structure'. Many societies (in Britain) and academies (on the Continent) were founded for the serious discussion of rural affairs, the exchange of information and the publication of papers; such were the Bath and Highland Societies in Britain, and the Economical Societies of Celle, Cassel and elsewhere in Germany, of Prague in Bohemia, and in France the Agricultural Society of Paris, with less active ones in Spain.

Britain was an early leader of this agricultural renaissance, following on Townshend and Tull's improvements in arable crop husbandry and the writings of Arthur Young (325–328) and others. The Bath and West Improvement Society was founded in 1777, and the Highland Society before 1787. These institutions attracted the support of many of the leading scientists of the day, such as Sir Joseph Banks, F.R.S. in England, men of affairs and political figures such as George III and Lord Somerville, and Sir John Sinclair in Scotland. Sinclair was the moving force behind the setting up in 1790 of the Board of Agriculture in London which, stimulated by the writings of Arthur Young from 1770 onwards, led to the compilation of the 'General Views of Agriculture', county by county, between 1790 and 1815. In Germany (289, 298) and France (64, 92, 93, 180, 297) there was intense interest in the production of fine wool, based on the importation of Spanish merino sheep, under the active support of the heads of state, especially in Saxony, Silesia, Prussia and in the neighbourhood of Paris. Wool and the sheep industry still remained an important element in the national economies, and warranted the personal interest of royal patrons and of Governments. It is this socio-economic background which endows with a special value the European literature of the later 18th and early 19th century on scrapie disease, a literature which, although largely lost sight of in the English-speaking world, remains alive in Germany (199, 130). The following summary reflects a rather rapid survey of certain important papers which are available in English libraries, together with a number of key German papers, for knowledge and photocopies of which I am deeply indebted to agrarian historians and veterinarians in Berlin, Budapest, Göttingen, Hanover, Hohenheim, Munich and Prague. For similar material from France and Spain I have received assistance from Toulouse, Barcelona and Madrid. It must be realised that this summary makes no claim to be complete, or to give more than brief accounts of books and articles referred to, some of which have had to be taken from secondary sources. A full reading would require work far beyond the scope of the present study. In the hope that the information may be of interest I feel it should be placed on record and not 'lost' again.

II. The Structure of Contemporary Sheep Populations

The available evidence suggests that the prevalence and indeed occurrence of scrapie is closely connected with the breed and genetic structures of the sheep population in which the affected animals are observed. The following brief account has been selected for its possible relevance to our understanding of the natural history of scrapie as it has been recorded in Europe, and especially England, and equally its non-occurrence in Australasia. For more detailed information the reader may consult the specialist works by Ryder (257, 258), Trow-Smith (306, 307) and others.

The structure of the domestic animal population is determined by many artificial constraints not operating in natural breeding populations. Apart from the nature of the indigenous foundation stock, the artificial mating programmes imposed by breeders since at least biblical times also involve controlled grazing and environmental factors. These have varied with the changing aims of flockowners and shepherds, and with the available grazing and housing, which in turn are influenced by agrarian and pastoral practices and economies, and have varied markedly from place to place and from century to century.

Over most of Western Europe, with the exception of Spain, the period since 1700 has been one of improved nutritional and environmental standards of living for sheep; it has also seen a marked increase in the intensity of phenotype improvement by genetic selection. This renewed interest in breeding began about 1730, when efforts to improve the performance of the English thoroughbred racehorse, based on Newmarket, broke away from the generally accepted safe low level of consanguinity, and adopted a system of closer inbreeding. The extreme example of this was termed 'breeding in-and-in', i.e. the mating of father to daughter, son to mother, brother to sister, uncle to niece etc. This system, or some similar form of close in-breeding, was adopted by Bakewell of Dishley, Ellman of Glynde and others; the extent of

Ellman's in-breeding was the cause of strong warnings from Arthur Young (116) but it allowed 'fixing' of a prepotent preferred genotype. All was well so long as the foundation stock carried few deleterious genetic traits; if any were present, the system of recording permitted their immediate recognition and elimination by radical culling, such as Ellman carried out. Others were less successful; Bates' Duchess Shorthorns developed an imperforate hymen with complete infertility, the cause of which was not recognised at the time. Similar 'breeding in-and-in' was adopted in the Spanish merino studs of Germany and France in the late 18th century. The results were catastrophic within 20 years, when scrapie became so prevalent that certain stud flocks virtually ceased, as the increases failed to replace their losses (199). With a disease such as scrapie, manifesting in middle age, and usually at least half-way through an animal's reproductive life, the control of any genetic predisposition becomes nigh impossible without individual identifications and meticulous mating and lambing records, not always easy to carry out under 18th century farming conditions.

III. Sheep Stocks in Western Europe

The following should be taken as a very tentative and imperfect account, the data required being scattered in many places, and nearly always found to be defective in some particulars relevant to our present purpose. The original archaic sheep stocks of north-western Europe prior to the Roman Empire are thought to have consisted of small (about 30 kg body weight) horned sheep with a hairy fleece, with many pigmented and medullated fibres, resembling the Soay, St. Kilda and Manx Laoghtain in Britain, the Tautar of Northern Scandinavia and the unimproved Iceland breeds of the present time. The human movements during the four centuries of the Roman Imperial occupation of Europe were almost certainly associated with movements of sheep, as a food source at the least. As the Romans occupied the Mediterranean littoral and Asia Minor, it is thought (257, 258) that there were successive introductions of sheep from the east and south-east, some hornless and some horned. By the end of the first millenium AD some black-faced horned sheep of astralgi type originating from the Black Sea and Caspian region may have reached the Baltic states and have been introduced by the Danish invaders into the Danelaw area of eastern

England, giving rise later to the horned Heath Sheep of northern Britain.

At the beginning of the second millenium AD the Viking invasions ensured an admixture of the sheep stocks of northern and western France into Britain. After 1200 AD the international organisation of the Christian monastic orders of the Roman Catholic Church led to renewed interest in livestock improvement, as monastic houses sought to increase the income from their estates in the face of inflation. In particular the Cistercians (174), with their late settlements after 1100 AD on poorer land at the margins of the Norman manorial occupation in Britain, for instance in West and North Yorkshire and Wales, found themselves farming ground not suitable for close divided units of settlement. They set up extensive sheep farming ranches, managed from the centre by means of outlying granges with the aid of lay brothers. Sheep farming was their main enterprise, and by the middle of the 13th century they had established a strong trading connection with the wool magnates of Flanders, Florence and Venice, and were the first to 'class' their wool in three grades. The Cistercians were thus able by their internal organisation and international connections to play an important role in improving the British sheep stocks; for example the Cistercian abbey of Whitby imported sheep from Spain through the port of Hull in about 1240 (143).

There are scattered accounts of specific movements of sheep by royal and monastic foundations in the Middle Ages, such as the consignment in 1274 of 1,200 sheep to England from Spain as a gift from the King of Spain, recorded in the St. Albans Chronicle (260). There are also oral traditions that before the Reformation merinos were brought into Scandinavia from Spain by certain monasteries (217). Although there was a strict embargo on the export of merinos from Spain by secular owners, there may have been some clandestine movements of merinos into Northern Europe prior to the recorded movements of the latter part of the 18th century (189). The increased value of, and demand for, fine wool in the 13th to 16th centuries would be a strong inducement to such movements (240).

In Britain by the 17th century two main groups, the white-faced and black-faced strains, had emerged by inter-breeding from the successive introductions of sheep, some hornless and some horned, of astralgi type (35, 257, 258, 11), while leaving untouched some indigenous hill breeds like the Herdwicks, which show affinities with the present native sheep of parts of Central Asia and the Western Himalayan and Pamir region. In Britain the white-faced fine-woolled hornless sheep gave rise to the short-woolled Ryelands or Arkenfield sheep of Herefordshire and the West Midland lowlands on the one hand, and the long-woolled Cotswold, Romney and Lincoln breeds on the other, while in the hill country of the north-east the dark-faced horned mountain breeds evolved, developed in the Danelaw area from the astralgi-type introductions.

The 18th century was a period of rapid development and fixation of breed types; in England, for example, the Southdown type established by Ellman of Glynde, the Leicester by Bakewell of Dishley, the New Kent by Goord of Iwade, and the Electoral merino in Germany. These were but instances of the massive effort in agricultural improvement and livestock breeding associated with the European 'Enlightenment', which received active encouragement from many royal and influential persons, notably in England and Germany. The improvement of wool became a major economic issue, with a shortage of supplies of English and Spanish fine wools (243, 240), derived in the main from Ryelands, Shetlands, and the undercoat of Wiltshire and Norfolk Horns in Britain, and the travelling Spanish merinos from the aristocratic *cabañas* of Spain, under the control of the Mesta (see Klein; 173).

Since it is from Western Europe that the present large sheep populations of the Americas, Australasia and Southern Africa were built up from the early 19th century onwards (65, 85, 13), a brief account of some salient particulars of the sheep population of Western Europe in 1800 becomes necessary before examining the occurrence of scrapie, since migrations of sheep convey genes even more certainly than microbes.

By 1800 in most West European countries many distinct breeds had become 'fixed', i.e. sufficiently homogeneous genetically to breed a recognised predictable phenotype. Twenty-five breeds were recognised in Britain, including Herdwicks (201). The numerically important breeds were the Dorsetshire (Dorset Horn), Wiltshire Horn, Cotswold, Southdown, and Hampshire Down in Wessex, with the Mendip, Portland, Bampton Notts and Berkshires less well defined. In East Anglia the dark-faced Norfolk Horns of Norfolk and Suffolk developed in the 17th century (248); in the south the hornless white-faced Cambridgeshire predominated, with the long-woolled Lincolnshires north of the Wash, and New Kents in north Kent and the Romneys on Romney Marsh. Further north the black-faced horned heath sheep (88, 35) were populating the hills from Northumberland and Lancashire to Fort William in Scotland, and by 1800, with the aid of the Bishop of Durham, had evolved the speckled-faced horned sheep of the Middle Pennines, the forebears of the Swaledales and Dalesbred, the Rough Fell, the Lonk and the Derbyshire Gritstone. The Teeswater breed in the lowlands of Durham and North Yorkshire was of a well-fixed large type with a fine-lustre silky wool, and prolific as it is today (35). The small Shetland sheep were a distinct short-tailed breed, highly valued for their wool (148). The Cheviot of South-country type was probably well characterised before 1800 (275, 36).

In Wessex the basic commercial sheep stocks in 1750–1780 were the Wiltshire Horn in the Salisbury Plain–Cotswold region, the Dorset Horn in southern Dorset and the Isle of Wight, the Hampshire in the eastern areas of Hampshire and Berkshire, with the heavier unselected Berkshire-type speckled-faced sheep in Berkshire and East Oxfordshire. The Dorset Horn was already a relatively 'fixed' type, with the Wiltshire Horn representing a more ancient sheep type akin to the Norfolk Horn. A small number of improved breeds were being selected —a selection that must have entailed detailed pedigree recording: the Dishley Leicesters by Bakewell (15, 103), the Southdowns by Ellman of Glynde in Sussex (111, 116), the New Kents by Goord of Iwade in north Kent, and the Cotswolds by crossing with Dishley Leicester rams. A small but increasing number of rams of these breeds became generally available from about 1780–1790 onwards, to which a few select Spanish merino rams may be added from 1787. No mention of scrapie in any of these four breeds has been encountered, and the English Leicesters, Southdowns, Kent or Romneys and Cotswolds are almost certainly free of the scrapie trait at the present time.

The Wiltshire Horns, Dorset Horns and Hampshires were probably bred on a ranching system with ram group matings, since the necessary enclosures were not yet available on the open downland (94, 95). Indeed the recording systems used in these breeds even in 1940 did not allow accurate pedigrees to be established. They were affected by scrapie in the mid-18th century, and they still are.

Cotswold, Dishley Leicester and Southdown rams were used from 1789. Spanish merino rams became available from 1792 onwards (68), when rams were donated to the Bath and West Society from George III's flock through Sir Joseph Banks and placed with members of the Society for group use, e.g. two with Mighell of Kennett, Marlborough in 1792, two with the Earl of Ailesbury at Savernake in 1795, while four merino rams went to Longleat, Somerset, for the Marquis of Bath and two to Cirencester, for the Earl Bathurst in 1792. These early merino introductions were followed up rapidly by many more between 1795 and 1810. In 1808 the Junta of the Asturias made a gift of 2000 sheep to George III and 214 to the British Cabinet, which were specially selected from eight Leon summer 'dehesas' (pastures) and shipped from Gijon on 4th October, 1808; 1800 ewes and 120 rams reached Portsmouth, of which Portland and Canning took 90 head, Castlereagh 30 and Mulgrave four, the remainder joining George III's flock at Richmond. These latter were joined in July 1809 by some 2000, shipped from Cadiz, of the remnants of the Negretti *cabaña*, which had been driven south into the Guadalquivir valley in search of new summer grazings to replace the traditional ones of Leon lost by war. There were other direct importations of merinos to Taunton, Somerset and to Devizes, Wiltshire; and dispersal sales from George III's flock between 1810 and 1820. These merinos were of Negretti stock, with some Paulars after 1809, and in Trinker's importation through Gijon in 1811. The breed structure of the flocks in the region changed remarkably between 1790 and 1810 (94, 95). Southdown rams were used extensively on Wiltshire Horns in the central area of Wiltshire. On the Marlborough downs, Mighell of Kennett used them first as an outcross on his flock from 1789, and changed his flock from Wiltshire Horns to pure Southdown type, with some admixture of Spanish Merino crosses. Others nearby, such as Black of Overton and Gray of Baycliffe, introduced some Southdown ewes and used Leicester rams on their mixed Wiltshire–Southdown flocks, as did Crook of Tytherton, Salisbury, while his neighbour, Hayter of Winterbourne Stoke, used Cotswold rams on his pure Wiltshire Horn ewes. Davis estimated that in 5 years there were 15,000 sheep of Southdown type in his area, and that Southdowns had largely replaced the Wiltshire Horns. They became so popular that in 1858 at Elston, near Shrewton, a very large pure-bred flock of 4180 prize Southdown breeding sheep were auctioned at one of the largest sheep auctions of a single flock ever to have been held in the South of England (216).

In the Cotswold area to the north, Cotswold rams were being used in 1790 by Hayward of Beverstone, Tetbury (308), and Sheppard of Uley, Stroud, had a large merino cross-bred flock by 1800 (68). From the turn of the century many commoners had merino or merino-crossed rams, for which numerous sources were available. Parsons (234) records the widespread use of merinos on commercial flocks in Wiltshire, Dorsetshire and Somerset, primarily to improve the wool but also because the merinos were considered to be free of 'goggles' (i.e. scrapie) and their crosses were likewise unaffected.

In Dorsetshire, Southdown rams were widely used on Dorset Horn ewes to control 'goggles' (77), while Balson of Athelhampton, Dorchester, replaced his Dorset Horn flock by Southdowns in 1800 for a similar reason (285). They were becoming widespread in Hampshire and the Isle of Wight, together with the speckled-faced Berkshires, by 1813 (313).

Dishley Leicester rams were being used in North Somerset (41) and by Bridge of Wenfold Edge (77), together with a merino ram. They were used from 1794 by Crook of Tytherton and others in South Wiltshire (95) for improving their flock productivity and reducing losses (e.g. from 'goggles').

IV. Present-day Breed Structure in Britain

The evidence for the history of the sheep breeds in Britain and their possible derivation has been reviewed by Trow-Smith (306, 307) and studied especially by Ryder (258). Although descriptions of British sheep types are almost non-existent until the 18th century, evidence from skeletons in archaeological sites, and wool characters from textiles, fabrics and parchment do allow plausible reconstructions of types. Armitage and Goodall (11) have published recent archaeological and iconographic evidence on the medieval Horned and Polled Sheep.

For our present purpose the British sheep population may be considered to be derived from four main sources.

(1) The ancient archaic stocks represented by the small brown-fleeced horned Soay sheep.

(2) White-faced mainly hornless stocks, thought to have come in with the Romans 2000 years ago, and from which the medieval short- and middle-woolled sheep were developed.

(3) A black-faced horned and hairy stock, with argali-like horns, thought to have originated from the Asiatic hairy sheep of Scythia, and introduced into the area of the Danelaw in north-eastern and eastern England by the Danish invaders of the 8th and 10th centuries AD.

(4) The Spanish fine-woolled sheep, or merinos, first introduced about 1260 (260), when 1200 were imported, followed by the numerous importations of the late 18th and early 19th centuries.

The present British breeds, largely evolved since 1700, may be categorised as sub-races or admixtures of these four main types (258). They comprise (1) the short-woolled breeds of Ryeland type, derived from the medieval hornless white-face, (2) the south-west horned type derived from the last-named with an infusion of archaic stock such as the Soay, (3) the medieval long-wools, giving rise to the Romney, Cotswold, Leicester, Lincoln and Teeswater breeds, (4) black-faced horned types derived from the Heath sheep (35) and giving rise to the Norfolk Horn, Scots Black-face and the speckled-faced horned breeds of the Pennines. The conventional modern classification recognises three main groups—short-woolled, long-woolled, and mountain breeds (Figs. 2.1, 2.2, 2.3). Merinos (Fig. 2.4) have long since ceased to be a standard British breed.

Each sub-race in turn is composed of phenotypically distinct self-contained breeding groups, of which there are 48 separately identified breeds, each with its own breed society or association maintaining a register of pure-bred flocks and the breeding stocks, at least of the rams (206). There are also at least 6 F1 inter-breed crosses of recognised half-breeds, to distinguish them from the many mixed cross-breeds of varying and uncertain ancestry. At present there is no practicable method of determining the parental genotype contributions to the phenotypes by blood typing or from studies of potassium and haemoglobin content of erythrocytes. Each breed is a closed breeding population of up to 1000 or sometimes more flocks of 5 to 2500 breeding ewes, many of under 50, organised under separate ownership and running on different estates. Within each breed it is usual for a small number of élite stud flocks, 2–5% of the whole membership, to breed the rams most sought after by members; these élite nuclear flocks thus exert an

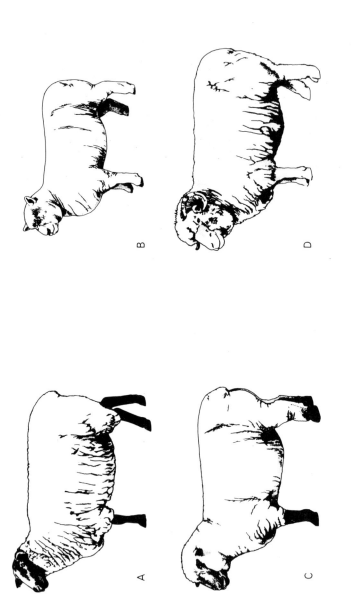

Fig. 2.1. Some British short-woolled breeds: A, Suffolk; B, Southdown; C, Hampshire Down; D, Dorset Horn.

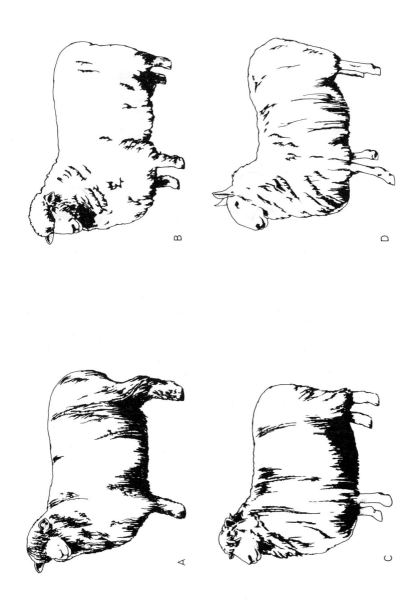

Fig. 2.2. *Some British long-woolled breeds: A, Lincoln Long-wool; B, Romney Marsh; C, English Leicester; D, Border Leicester.*

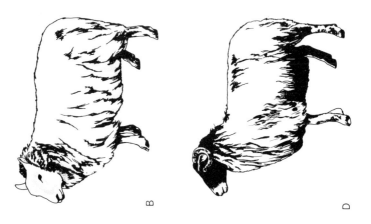

Fig. 2.3. *Some British hill breeds : A, Welsh Mountain ; B, Cheviot ; C, Scottish Black-face ; D, Swaledale.*

Fig. 2.4. *A South African Merino.*

influence on breed structure quite disproportionate to their numerical size, and hence demand special attention.

Of the 48 breeds many have become so reduced in numbers as to be barely viable genetically; only some 15 are the main contributors to the British national flock at the present time, and wherever possible it has been these breeds and the élite stud flocks of these breeds which I have studied (see Chapter 5). Of the principal breeds, there is documentary evidence that many were well-established in the 18th century (e.g. Norfolk Horn, Wiltshire Horn, Dorset Horn and Hampshire Down, which were all affected with scrapie, and the Southdown, Ryeland, Lincoln, Romney, Cotswold and Leicester, which were not); while several others, e.g. Cheviots, Teeswaters and the long-woolled hill breeds of the Pennines and Scotland, were certainly recognised by 1800 (275, 36), although evidence of scrapie is lacking until later (324, 286) (Table 2.1).

The 19th century saw the emergence of many new breeds (282, 65) produced by varying admixtures of the older breeds, and by selection of locally important previously unselected stock, e.g. the Clun Forest and Kerry Hill and West Country breeds. The Cotswold on the Hampshire Down gave rise to the Oxford Down, while the Southdown on the Hampshire Down produced the Dorset Down. Neither of these new breeds has been found to carry the scrapie trait. One parental

Table 2.1. *The principal sheep breeds in Britain*

Breed	Area	Approximate date	Origin	Affected by Scrapie				
				1750–1800	1800–1850	1850–1900	1900–1950	From 1950
Cotswold	North Cotswolds	before 1700	Very ancient; fine wool	+++	++			+
Romney	Kent and Romney Marsh	before 1700	Ancient, lustre wool	+++	++			++
Norfolk Horn	East Anglian Heights	before 1700	Selection from lowland heath sheep					++
Wiltshire Horn	Wessex	before 1700	(Declining breed)					
Welsh Mountain	Welsh Hills	before 1700	Several local types					
Herdwick	Lake District Fells	before 1700	Ancient hill type, like Icelandic and Central Asian					++
Arkenfield or Ryeland	Herefordshire	before 1700	Selection for fine wools	+++				
Dorset Horn	Dorset and Isle of Wight	before 1750	(Polls introduced since 1950 from Australia and via Ryelands)		+			+
Lincoln Longwool	Lincolnshire and East Yorkshire	before 1750	From local old longwool type with some Dishley blood					
Teeswater	Durham lowlands	1700–1750	Well established by 1791 (35) with twinning					++
Southdown (Ellman)	Glynde, Sussex	1750–1810	Phenotype selection and close in-breeding					
Speckled and black-faced heath	North Pennines to Fort William	1750 on	See Bewick (35)					
Dishley Leicester (Bakewell)	North Midlands	1760–1810	Phenotype selection with progeny testing					
Cheviot (South Country)	Cheviot and South Scottish hills	1760–1810	Uncertain (36)			+	++	++
Border Leicester (Culley and others)	Tweedside and Northumbria	1780–1860	Selection and ? Cheviot infusion			+	+	+
Shetland	Shetland Isles	before 1790	Fine wool breed					
Scottish blackface	Scottish hills	1790–1850	Selected for face colour from heath sheep				++	+

Breed	Location	Date	Origin			
Swaledale	Durham and middle Pennines	1790–1850	Selected for lighter speckle from heath sheep			+++
Hampshire Down	Salisbury Plain	1790–1820	Southdown ♂ × old Hampshire and Berks. ♀			++
New Kent (Goord)	North Downs	1790–1850	Selection from old Romney stock	+++		
North Country Cheviot (Sinclair)	Caithness and Sutherland	1795–1820	Merino (Negretti) ♂ on SC Cheviot ♀	+		
Oxford Down	Cotswolds	1800–1850	Cotswold ♂ × Hampshire Down ♀		+	
Suffolk	East Anglia, especially Suffolk	1800–1850	Southdown ♂ × Norfolk Horn ♀		++	+++
Shropshire	South Shropshire red soils	by 1810	Probably Ryelands ♂ on Down type ewes			
Kerry Hill	Knighton area of mid-Wales	1810–1850	Local selection of speckled-faced breeds			?+
English Leicester	East Wolds of Yorks and Lincoln	1820 on	Some infusion of Lincoln ♂			
Wensleydale	Wensleydale and North Lancashire	1830–1850	? Leicester on old Yorkshire Longwool			
Lonk	South Pennines on limestone	1830–1880	Local selection of Swaledale type			
Derbyshire Gritstone	Derbyshire and around Sheffield on gritstone rock	1830–1880	Local selection of Swaledale type			
Dorset Down	East Dorset Downs	1843–1860	Southdown ♂ × Hampshire Down ♀			
Border Leicester × Cheviot	Yetholm area	before 1850	Fixed half-bred from older Cheviot ewes	++		
Blue-faced Leicester	Hexham district and Cumbria	1850–1900	Culley's Border Leicester × Wensleydale			
Clun Forest	Clun Hills	by 1875	Local selection of brown-faced breeds			+
Dalesbred	Wharfedale	1930	Selection from Swaledale type			?+ +++

stock, the Hampshire Down, carried the trait, but the Cotswold did not. The scrapie-affected Norfolk Horns were mated with Southdown rams to give rise to the modern Suffolk, which carries the trait; the Norfolk Horn crosses with Cotswold and Leicester rams have not led to any recognised breeds today. The English Leicester, which was and is probably still today free of the scrapie trait, gave rise to the Border Leicester, which now carries the trait, as does its off-shoot, the Blue-faced Leicester. For the Cheviot, evidence suggests that the trait was manifest by 1850.

Evidence concerning the Northern and Welsh Mountain breeds is fragmentary, although they were probably stabilised genetically by the early 1800's (36); we have found no documentary evidence for their carrying the scrapie trait prior to recent times. However, in spite of assurances that certain breeds were free in the early 1950's—assurances upon which experimental scrapie studies were based—it soon became clear from our own observations that at least two, the Welsh Mountain and Swaledale, carried the trait. This has since been confirmed by the observations of others (236, 58) as well as our own. How and when these hill breeds and the Teeswater came to acquire the trait is uncertain.

V. The Spanish Merino and its Diaspora, 1760–1815

Of crucial importance to the sheep populations of Europe was the movement of fine-woolled merino sheep out of Spain, although their export was officially prohibited (28). Until the 18th century, probably few merino sheep left Spain. After 1700, however, many consignments reached northern Europe, some of which were associated with severe outbreaks of scrapie disease and some of which were completely unaffected. Thus the first recorded occurrences of scrapie disease in Germany (181), in France (129), and in the Danube Basin of Central Europe (169, 254) were associated with, and frequently attributed to, the importations of the Spanish merino, among whose descendants the disease occurred in epidemic form in Germany and France (199). However, the importations to Britain and Scandinavia did not show the disease, and indeed in England between 1795 and 1810 the merino was used as an out-cross to control scrapie (the 'goggles') in indigenous breeds. In Hungary the original merinos were probably little if at all affected, and it was only after 40–50 years that the disease was reported

(9, 137, 138) (see p. 47). Furthermore, scrapie has not been recorded in the merinos and their many descendants in Australia, New Zealand, South Africa, South and North America, although some of these were derived from European stocks affected at the time of the original importations. Why not? The question calls for careful examination.

The Spanish fine-woolled merino sheep was a product of, and the key element in, the agrarian economy of the semi-arid regions of the Iberian peninsula for over five centuries from AD 1273 (173, 46, 28), its wool essential to the wealth of Spain and the political power of its kings, and its organisation protected by the greatest economic corporation in Spain, the 'Mesta', based on Leon in Castile. From the 14th century onwards the Mesta exercised a monopoly control, which remained intact until the 1830's (173, 66, 120, 189, 278). The merino flocks were transhumate, i.e. travelling continuously through non-settled unenclosed country under the care of a group of men with their dogs but without their families. They travelled slowly along established stock routes, which were usually at least 180 m wide (90 paces on each side of the highway), extending to several kilometres. For much of the year these flocks covered as much as 800 km between their traditional summer grazings in the higher north central parts of the Central Plateau (289). They were always kept in the open and not housed, nor were they fed conserved fodder. The environment of these central areas was one of treeless, sparse semi-desert or scrub without enclosures or shade, with hot days and cool or cold nights, and mostly with little rain; indeed, the present annual rainfall is 10–15 inches, and in parts of Murcia and Cuenca, rain may be absent for 8–9 months at a time (119, 66). In 1787–1788 Beckford of Fonthill could ride on horse-back for 5 hours across this desert-like country without seeing a human being or meeting a man-made obstacle to his progress. Sheep acclimatised over 500 years, or at least 100 generations, to such an environment developed an ability to walk long distances in obtaining grazing on a modest water intake, and became accustomed to grazing mature high-dry-matter forages of low nutritional value. As over-grazing and soil erosion increased over the centuries (66), the fineness of the wool was no doubt improved, but the ewes had no more than a single lamb, and the rams were not used for mating until the age of 3 years (289). Shearing took place in early summer in the neighbourhoods of the four 'wool and sheep' towns of Leon, Soria, Segovia and Cuenca, upon which the international wool trade was based. Sheep were mated in late summer, when for 3 months they had a special allowance of salt before beginning their journey south to their winter grazing, where they lambed in December to February, so that their lambs were 2–4 months old, and strong enough to travel north, in April. There was another special characteristic of this merino sheep industry. The shepherds of migratory (transhumate) flocks were socially a world apart in the countryside, leading a peculiarly isolated life away from permanent settlements, and were cut off by distance and temperament from normal relationships with the local settled peoples. This rift developed into a rigid caste system of 'untouchability', shown by a special status barrier towards the shepherds as social outcasts (46), which had persisted from the 16th century (316) until this century (120, 122). Indeed this special separate world of the shepherds with transhumate flocks persisted into the 20th century in southern Germany (163) and in Provence (198). Furthermore, in 18th century Spain personal visits by an owner to his estate, let alone to his *cabañas*, were so rare that they were accompanied by special celebrations (66).

Under these circumstances it is perhaps not surprising that accounts of husbandry and management by flock owners such as are found in England and Germany are virtually absent from Spain. By far the most comprehensive details of reproduction performances, general husbandry and health records have been found in a remarkable report by Stumpf of Leipzig (289) based on the manuscript memoirs of an anonymous native of Dresden who had visited Spain for some time, probably in the 1770's, had proved a congenial visitor, and was obviously well informed on the shepherding practices throughout the year of the Leonese *cabañas*. They had changed but little a century later (120); but today the migratory merinos have very largely disappeared (257).

Many of the technical terms used in merino flocks are Arabic, e.g. *morrueco* for breeding ram. By the 15th century the Spanish merino sheep, a cross between Spanish native (*churro*) sheep, with their reddish wool, and the North African fine-woolled sheep introduced by the Mohammedan conquerors after 752 AD, produced wool of the highest quality (189); by the 18th century, this wool was produced only by the transhumate merinos, and not by the non-migratory merinos (189). The word 'merino', which has been the subject of much specula-tion (28), appears to be derived from the Banu-Marin, or Merinids, a desert tribe of Southern Morocco, and was applied to a very special grade of fine wool shipped as early as 1307 from Tunis to Genoa, where it was woven into a very high grade cloth which could compete with the best imported English cloth (189).

By the 18th century Spanish merinos were very carefully organised in a small number of large detachments or *cabañas*, some of 30,000–40,000 head divided into 40–50 separate groups with their own night-folds, or *dehesadas*, under the separate ownership of territorial grandees, military orders and church foundations, less than 20 in number, who were prohibited by royal decree from exporting, and indeed almost never selling except under duress, e.g. the sale of Paulars to Manuel Godoy in 1796 (68). Occasional gifts and exchanges might be arranged through diplomatic channels; for instance, the despatch of 37 Paular ewes and four rams from the Count and Countess del Campo di Alanje to George III of England in 1791 in return 'for 8 bay (? Cleveland) carriage horses of uncommon beauty'(68). Each *cabaña* had sharply defined main stock routes (*cañadas*) and static grazing grounds (*dehesadas*), and their head shepherds were men of considerable standing in their communities (173, 120). These *cabañas* represent distinct breeds, as the term is used in Europe at the present time, or 'types' in the Australian sense, of the Merino race (13), and differed as much from one another as modern British short- and long-woolled breeds. Thus in discussing the precise role of the merino in the dissemination of scrapie, one requires to know the *cabañas* from which the population is descended. Such knowledge is hard to come by.

The average numbers of sheep in the Mesta flocks between 1477 and 1563 ranged from 3·45 million in 1526 to 1·67 million in 1562; and in 1910 they accounted for 1·5 million out of a total of 14 million sheep in Spain (173). The rules of the Mesta were strictly applied until 1834 (8, 173, 120). The breeding regulations, derived from Moorish Berber practices (256) and codified by Lapata in 1820, were in practice entirely in the hands of the *mayorales* (head shepherds) without the close supervision of enlightened owners like Fink or Ellman. It ensured careful matings in groups (*rebaños*) of 400–1000 ewes with 5% of rams, all older than 3 years, which were held on separate areas (*dehesadas*) of open ground demarcated by flat stones (289, 180, 68, 28). Thus each *rebaño* represented a closely-bred clan, often referred to by the name of its *dehesada*, without individual pedigrees. The system thus provided clans of separate blood-lines, without close inbreeding, within a main *cabaña* of 10,000 or more sheep, under the control of a single *mayoral*, who in the case of the Paulars in 1810 could formally testify before legal authorities that no admixture of outside blood from another *cabaña* had occurred during the 24 years of his stewardship (68).

In the 18th century a special mark of Spanish royal favour was the gift of a consignment of transhumate merino sheep. Only rarely did other merinos leave Spain, except as unrecorded contraband, mainly through Portugal (68). Consignments, usually transported by sea to North Sea ports, arrived in Saxony in 1713 and 1740 (273, 289), and became widely distributed in the electoral and farmers' flocks of Saxony and Anhalt-Dessau, where they found a ready outlet for work in the large marketing centres of Dresden and Breslau. A royal introduction reached Sweden in 1723 and formed the flock at Alsingfors in 1739 (264). A small flock was established in France at Alfort near Paris in 1764, and a larger one at Rambouillet in 1786, by a royal gift to Louis XVI of 300 sheep from 10 of the best élite *cabañas* of Spain (68). In Germany a gift to the Elector of Saxony came in 1765 to form the main foundation of the royal stud flock at Stolpen. Small numbers came to England in the 1780's before George III's importation of 1788, to be followed by numerous others before 1799 (68).

The movement of merinos into central and south-eastern Europe was less easy, being usually by sea to Fiume and thence on foot over the Alps into the Danube basin, with losses of 50% of the consignment. Spanish merinos were introduced into Hungary from the middle of the 18th century, the Empress Maria Theresa making a special importation in 1773 (169). From 1778 the merinos went mainly to the Imperial flock at Holditch (282) and to the estates of aristocratic landowners (137), where some were bred pure, but many were used for cross-breeding with the indigenous sheep. Most of these merinos were of the Escorial race (6, 331). The Imperial flock in Austria was established in 1786 (282, 130).

By 1800 the Spanish merino was present over much of north and central Europe from Hungary to Scotland (275), later as far as the Shetland Isles (68), and was being used extensively as an 'improver' ram on native sheep (289, 118, 298). Between 1800 and 1815, with the disruptions of the Napoleonic Wars and especially the Spanish War of Independence of 1808–1814, many merinos, particularly those of north-western and western Spain, were dispersed out of Spain, and became available to non-aristocratic flock owners (68).

With the military campaigns of 1807–1808 in the Iberian Peninsula, established winter grazing routes of many *cabañas* became disrupted, and many Spanish merino sheep of the Negretti, Paular and Escorial *cabañas* were purchased by English, Scottish and American buyers through Lisbon and Gijon (68) those coming to England being mostly Negrettis (302) with Paulars going to Trinker of Devizes and to Downie of Paisley in 1810 (105). A merino ram reached Norway in 1750, but there were very few before (1, 211, 217). Merino rams were auctioned at Viborg in Denmark in 1809 (318), and 50 rams were sold to Norway in 1811 from Esrom, Denmark (211, 217). From 1810 Spanish merino rams were widely available in Northern Europe for cross-breeding, as in Britain, but the number of pure-bred merino flocks outside of Saxony, Silesia and Bavaria remained small, and their numbers gradually declined over the 19th century in Britain (68), as in Scandinavia (207). This was probably primarily due to climatic conditions affecting wool quality, and competition from Australian wools, produced under more favourable climatic conditions. The German Saxony merino and the French Rambouillet merino survived as distinct and important breeds into the 20th century. The German Merinolandschaf still remains as migratory flocks in the hill country of south-western Germany, while the settled Merinofleischschaf occurs in the north around Hanover (319).

The precise origins of the Australian merino are known only in broad outline. Small numbers of Spanish merinos, obtained in 1797 from the Cape of Good Hope, are said to have been of Escorial stock derived from the Netherlands (13, 28). A small consignment from George III's English flock of 'old' Negretti stock went to Macarthur's Camden Park flock in New South Wales in 1805, to which were added in 1812–1815 a consignment of Rambouillet sheep from the Australian Agricultural Company (68). Meanwhile, many undocumented consignments of the Estremadura transhumates were made from Lisbon between 1802 and 1827 in the aftermath of war (261), probably mostly Negrettis with some Paulars. From Germany importations of Saxony merinos, mainly of Escorial breed, but with some Negretti, Paular and Infantado sheep, were made before 1830.

Following the droughts of the mid-1850's in Victoria, substantial importations were made of new merino 'blood' in the form of imported rams, which were exhibited at Sydney and Melbourne sheep shows.

Silesian Electoral stock came from Klippenhausen and Leutewitz as well as rams from Rambouillet, with some European Negretti rams, while many Vermont merinos were imported from the U.S.A. between 1866 and 1890 (13). Detailed information on the composition of the foundation stocks of the Australian merino population is incomplete and scattered. The earlier data have been extracted by Carter (68: Plate VIII, and Figs 7 and 8). The most authoritative stud breeder's account is that of Austin, of Wanganella (13), which may be supplemented by those of Cox (85), Guthrie (139) and Belschner (29). The contribution of the original individual Spanish *cabañas* to the Australian merino population must remain uncertain. The main importations prior to 1820 were predominantly of Negretti and Paular blood, with some Escorials, probably free of the scrapie trait. The mid-century importations from Rambouillet and of Saxon Electorals from Silesia were made at a time when scrapie had become virtually unknown, and no record of its occurrence in Vermont merinos in the United States has been found. Thus the later importations are unlikely to have been in contact with scrapie; some of the earlier Escorials may have been. There is no evidence of the occurrence of the disease, let alone its establishment, in Australia or in New Zealand, which received most of its merinos from Victoria, Australia, in the mid-1950's (55).

Following this account, unavoidably sketchy and incomplete, of the main sheep populations of Europe and of their movements in the last few centuries, we are in a better position to assess the significance of the various outbreaks of scrapie during the same period. This is the subject of the next chapter.

Chapter 3

Recorded Occurrences of Scrapie from 1750

There is an extremely rich and competent literature on scrapie in the English and German agricultural writings of the later 18th and early 19th centuries, written for the most part by men with first-hand practical knowledge of the disease. These writings leave little doubt as to the correctness of the diagnosis. Regrettably the recent reviews of scrapie most readily available in English make little or no reference to this source of epidemiological data. It seems desirable, therefore, to set out in some detail, incomplete as this must be, information which scholars in many parts of Europe have so kindly drawn to my attention, and to provide a supporting bibliography, much of which is little known and inaccessible in Britain. Indeed the divorce in British agricultural education of 'Animal Breeding and Improvement' from 'Veterinary Diseases' (now called 'Animal Health'), which has not occurred in the same manner in Germany and Scandinavia, appears to be largely responsible for this lack of readily available information in Britain.

The data available for southern England during the period 1750–1850 are remarkable for their comprehensiveness, and the general high level of openness and reliability. The German summary of continental studies by May (199), given in English translation by M'Gowan (204), has proved to be even more comprehensive. There is in addition a substantial literature on scrapie in Eastern Germany and Silesia between 1785 and 1850, and also in Hungary. These writings support the English evidence in all material respects.

On the other hand, we rarely find 'follow-up' studies covering a decade or more; reports tend to deal only with the earlier manifestations of major outbreaks. Moreover, no detailed accounts of precise breeding policies and mating schedules, or of the age composition of flocks, have been found. Often details of the breed and the quality of the sheep, stud or flock, are omitted. There are frequent reports in the English and German literature that the ram is a critical element in the causation of the disease, and that the use of rams from flocks free of the complaint will control its presence; but such freedom is very difficult to ascertain, since owners and shepherds will go to great lengths to hide the occurrence of the disease from enquirers, as happens, of course, today in Britain. To attempt to establish freedom from scrapie in a flock or an animal on the basis of verbal enquiries would be a hazardous undertaking, then as now.

The recorded occurrences of the disorder are hence not a reliable guide to actual occurrence, but they do draw attention to markedly different frequencies in different breeds, and in different strains within a breed. It is important to know the structure of the sheep population from which the cases of scrapie have been collected, with the breeds, age composition and if possible pedigrees, especially of the rams used for the last two or three generations, together with the health records of near relatives; but such data are very rarely available. The summary that follows merely notes some occurrences of the disease, the totality of which may be likened to an iceberg, of which these recorded cases represent the tips visible above the sea-surface.

I. Recorded Outbreaks in Europe

Before 1700 it is not possible to disentangle the accounts of the various 'murrains' of sheep in Britain (306) and make precise diagnostic assignments: the older European literature has not been systematically searched, but is generally insufficiently detailed for diagnosis. After 1750, however, there are many excellent accounts in both German and English journals, giving small details which clearly confirm the identity of the disorder they are discussing for any reader acquainted with natural scrapie as it occurs today (Table 3.1).

It appears that scrapie, under the name 'rickets', was present about 1730 in England (82). By the mid-18th century the disease was sufficiently prevalent to constitute a serious economic constraint on the sheep industry in eastern England, especially southern Lincolnshire and Huntingdonshire, where sheep farmers petitioned the British House of Commons in 1754 to institute legal restrictions on those

Table 3.1. *Estimated prevalence of scrapie in Europe, 1700–1980, with references*

Date	England and Wales	Scotland	France	Germany	Danube Valley
1700–1750	1730: 'Rickets' observed in East Anglia (82)	Not recorded	Not recorded	Probably present (181, 115, 130, 27, 104, 215)	Not recorded
1750–1780	'Rickets' or 'Goggles' severe in East Anglia, especially in Norfolk Horns (167), and in Wessex in Dorset Horns, Hampshire and Wiltshire Horns (10, 95, 77, 308, 326)	Not recorded	'Vertige' (? scrapie) in Berry (178)	1759: observed in Saxony and Silesia, in merinos imported from Spain	Not recorded
1780–1820	Very severe in East Anglia and Wessex, as above, later declining (285, 312, 313, 41)	Not recorded	Becoming severe in merinos at Rambouillet (297, 129, 199)	Epidemics at Frankenfelde and Stolpen (259, 118, 199, 7, 299, 300)	Probably rare (254)
1820–1880	Marked decline with use of outcross rams (131, 158, 78)	Probably present by 1850 in Border Leicester and Cheviot crosses (63, 286, 287)	Moderate, but wide-spread (129, 253, 199). Milk sheep affected.	Severe in Electoral merinos in east, but not in south (199)	Severe in Electoral Merinos (9, 245, 330, 331, 169, 109)
1880–1910	Virtually unknown	Increasing in south (63, 204)	Sporadic outbreaks	Marked decline	Marked decline (160)
1910–1950	In East Anglia from c. 1920, mainly in Suffolks	Severe in South, affecting Cheviots, Border Leicesters, and Blackfaced and Halfbreds (202, 204, 286, 123, 135)	Sporadic in many areas. Pré-alpes and Berrichons affected (33, 191)	Becoming rare in Saxony by 1914. Very rare after 1920	None after 1930 (293)
1950–1980	Mainly in Suffolks; recorded outbreaks in Swaledales (58) Welsh Mountain (236) and other breeds	Sporadic outbreaks in Cheviots (323, 283) Dorset Down (340, 341) and others	Outbreak in Provence (165) in Pré-alpes breed	Outbreak in merinos in East Germany (147)	Outbreaks in Hungary (4) and Bulgaria (161) in Suffolks imported from U.K.

dealing in sheep purchased from breeders. No action was taken. Two centuries later, in 1958, the Suffolk Sheep Society, with encouragement, implemented the principal proposal of 1754 in its regulations for pedigree sales.

The disease was also serious in parts of central Europe by 1750 (181, 115, 199). Leopoldt (181), in recording *Traberkrankheit* in central Europe in 1750, specifically states that the disease occurred in Spanish merinos. The disease termed *das Drehen oder Traben* (i.e. 'turning' or 'trotting') was almost certainly scrapie (289). Stumpf (289) comments that the migratory shepherds of the merino flocks recognised it as a slowly fatal and incurable disease, which they detected at a very early stage, when they at once killed the animals for their flesh 'before they had lost bodily condition' exactly as is done today with scrapied sheep by knowledgeable owners. The disease appears to have been well known in many migratory merino *cabañas* (flocks) in Spain well before 1780, but was not found in the common non-travelling sheep. A similar disorder called *le vertige* (translated as 'dizziness') was serious in the French province of Berry (178). The term *vertigo* was still being used in Germany for *Traberkrankheit* in 1817 (299), and in France as a general term for the scrapie-*Traberkrankheit* group of disorders as late as 1827 (259).

Between 1750 and 1820 scrapie caused increasing concern in north-western Europe, occurring widely and sporadically, at times reaching calamitous epidemic proportions. In England (326, 95) the disease was clearly associated with certain indigenous breeds (the Dorset Horn, Wiltshire Horn and Norfolk Horn). On the continent of Europe, in Saxony (118), Prussia, Silesia and Moravia (7, 259, 113, 199) and in France at Rambouillet and elsewhere (129, 199) the outbreaks were said to be largely confined to certain strains of recently imported Spanish merino sheep, especially those from certain *cabañas* such as the Escorial, and notably those forming the fine-woolled small-boned Electoral breed of Saxony and eastern Germany.

Scrapie in Britain 1750–1820

The available literature contains many detailed comments, which leave little doubt that most writers were well-informed and familiar with the characteristic clinical behavioural signs peculiar to scrapie. It is beyond reasonable doubt that the disease they describe as 'goggles' or 'rickets' was not being confused with other complaints. Information is available for two main regions, Wessex (south-central England) and East Anglia (eastern England from the River Thames to the Humber), where 'goggles' presented a very serious problem to the sheep industry.

In Wessex the disease known there as 'goggles' had become so widespread by the 1770's that the first communication on livestock published by the Agricultural Improvement Society at Bath (now the Bath and West Society), founded in 1777, was on 'goggles' in Wiltshire (10), where the disease 'within these few years has destroyed some in every flock around the County and made great havock in many'. It was unknown in the area a generation earlier (81); i.e. the epidemic began about 1750–1760, and was severe from 1770 to 1810, when it declined (95) and was uncommon by 1820 onwards. No reference to the disease in the later 19th century has been found, and the name 'goggles' was unknown in mid-Wiltshire among experienced sheepmen in the 1920's, and until very recently.

'Goggles' was serious in the Dorset Horn breed by the 1780's (77), when the present breed was already well established, with its specialised autumn lambing system. It was even more serious in the spring-lambing Wiltshire Horn breed which at that time was the main producer of commercial flock sheep, females for breeding and wethers for wool and meat, in the Salisbury Plain–Cotswold region, from where sheep were sold and driven as far afield as Kent (308, 94, 95, 16).

By 1810–1815 the epidemic was abating. In Dorset it was less prevalent (285); in Hampshire it was no longer 'generally complained of' (313); in Somerset and Devon it was of little importance (41, 312) although to some it was still in 1800 'a most prevalent and dreadful disorder' (234). It was unremarked in Oxfordshire in 1813 (328), but memories of the disease lingered in Berkshire as late as 1840 (158), although it was not a problem at that time (78). The disease was not remarked upon at Woburn in Bedfordshire in 1795 (24) where South-downs were being used with Wiltshire Horn sheep.

It is clear from the records that scrapie was a major concern of many flockowners in Wessex and parts of East Anglia, and that deaths from the disease ran to many thousands (95), and was the main cause for the decline and virtual disappearance of the Wiltshire Horn and Norfolk Horn breeds. The Wiltshire Horn, Norfolk Horn and Dorset breeds were affected severely, and the Hampshire and Berkshire speckled-face breeds possibly to a lesser extent (326). All reports have indicated that scrapie was unknown in certain breeds—the Southdown, Cotswold, Dishley Leicester, the English-bred merino, Ryeland, Shropshire, Mendip, Portland, Lincoln and Kent (Romney Marsh)—and in their crosses (95, 308, 326). Details are not sufficient to say precisely whether the freedom in the F_1 generation continued in subsequent generations without continuous top crosses by rams of these unaffected breeds.

Scrapie was controlled, quite deliberately, by using rams of these unaffected breeds on ewes of the affected breeds. Evidence that selection of unaffected strains within an affected breed or breeding population was ever successful in controlling scrapie, although looked for, has not been found, and in the light of our present knowledge of the difficulties of identifying rams free of the scrapie trait in an affected breed (226, 228) any attempts are unlikely to have been successful, given the breeding and recording methods commonly available at the time. The use of rams of unaffected breeds obviously offered a simpler, quicker and more economical method of control, and was encouraged by the change from a ranching/herding to a folded sheep economy (94, 95), for which the new breeds were more suitable. Although some writers considered the disease infectious, no convincing evidence to support this view can be identified. It ran counter to the widespread recognition of the importance of breeding policies and ram selection in determining the occurrence of the disorder and its disappearance following the use of ram out-crosses of breeds not affected with the scrapie trait.

There were some alterations of breed goals, which may have favoured the selection of scrapie-prone animals between 1770 and 1790. Several observers comment on the concurrence of 'goggles' with a change of phenotype selection towards a larger, earlier maturing, more muscular animal, in the improved Wiltshire Horn breeds (94), in the Dorset Horn (77, 234), and in the Hampshire, by use of the Berkshire (234). Selection for early skeletal muscle development and compact conformation, as in the Suffolk and other breeds in recent years (228), has inadvertently led to selection pressures favouring animals carrying the scrapie trait. This phenomenon has been remarked upon in Germany and in Hungary (109).

In East Anglia the disease known as 'rickets' was recognised in Huntingdonshire between 1730 and 1760. In the decade 1745–1755 it was already an economic problem of such magnitude to sheep breeders in South Lincolnshire that public meetings held at Holbeach and Boston, Lincs., on 30th September and 23rd October, 1754, drafted a petition to the House of Commons which was considered on 15th January, 1755 (167). No official action was taken. Little further information is forthcoming until the end of the century, when the 'General Views of the Agriculture' in the counties were compiled. At that time the main commercial sheep stocks were of Norfolk Horn stock on the higher ground south of the Wash (131, 248), with Lincolns and Leicester crosses in the Fens, which had been opened to improved grazings by the drainage of the South Fen in 1702, and the Lincoln long-wool sheep stocks on the northern wolds of Lincolnshire, a breed well established and fixed by 1750 (82).

In the East Anglian Uplands from the Huntingdonshire–Cambridgeshire border through the Newmarket area to Bury St. Edmunds and northward to North Norfolk and South Lincolnshire, where the Norfolk Horn Breed had been developed and widely used since the early 1600's (248) the sheep were still predominantly Norfolk Horns (311a). 'Goggles' or 'rubbers' was then very serious, and a growing problem in individual flocks. In the Newmarket area 240 animals were lost out of a flock of 500 at Dalham (311a, 131). In Norfolk Horn flocks at Ashley and Silvery the disease 'was a growing menace, becoming as alarming as the rot' (fascioliasis) (311a, 131). In 1800 Young, who farmed in the district of Bradfield near Bury St. Edmunds, reported 'rubbers' proceeding to death, 'by which (distemper) some very capital flocks in the vicinity of Bury have lost several hundreds' (327). The description given is of classical scrapie as it has been observed in the same district since 1950. The disease was well-known in Norfolk about 1800 (322) where the Holkham Norfolk Horn flock was sold and replaced by Southdowns shortly after 1812 (248). In the South Lincolnshire fenlands, where the 'rickets' had been so serious in 1754 (167), the 'rubbers' remained one of the serious distempers of the marsh in 1813 (270), but was apparently of little or no concern on the Wolds to the north, where the Lincoln breed was kept. The last important Norfolk Horn flock at Lackford, near Newmarket, still had the disease about 1940; a few animals of this flock were seen about 1950, and an animal of the more ancient Norfolk Horn type, with a brown fleece, used in a television programme in 1978, exhibited early clinical signs of scrapie.

In Cambridgeshire, where the first pure-bred flock was established at Chippenham, near Ely, in 1792, Southdowns were soon being used as a top-cross on the commercial Norfolk Horn, West Country and Cambridgeshire ewes of the higher ground, among which scrapie caused annual loss rates of 40% in some Norfolk Horn flocks (311a). In the fens the sheep were more mixed with Leicester and Lincoln strains, and scrapie was not remarked upon. By 1811 the use of South-down rams had become widespread (131) and the breed, as in Wilt-shire, gained wide acceptance during the next 50 years. It was associ-ated with the famous flock of Silas Webb at Babraham, which was dispersed in 1861 and 1862. Such was the demand for these sheep that its first prize ram made the very substantial price of 250 guineas at Canterbury in 1860 (216).

In Lincolnshire, 'rubbers' was still a serious cause of loss in the fens in 1810 (95), but was not complained of in the wold flocks of long-woolled sheep based on Lincolns and their crosses.

In Norfolk, Suffolk and north Essex, the Norfolk Horns were increasingly mated with Southdown rams, to give rise by about 1850 to the distinctive Suffolk breed, which still carried the scrapie trait. On the lighter soils Cotswold rams were being used on commercial ewes from at least the mid-century (301) and were still widely used in north Norfolk a century later on ewes of the Suffolk breed. One of the advantages of the Cotswold cross progeny was freedom from scrapie. Merino rams were also used, but documentary records are less numerous than for Wessex. Sir Joseph Banks had a sub-flock of Merino × Sher-wood Forest sheep from 1788, and in 1790 mated two Norfolk Horn ewes from Holkham with a pure merino ram to provide a merino out-cross for the Holkham flock (68). In 1808 a merino flock based on Paular sheep was established by Western at Felix Hall, north Essex; at the same time merino studs based on Negretti sheep were established by Thompson at Nottingham, and by Sturgeon at South Ockenden near Grays, south Essex. These studs were dispersed in 1895 (68).

Coinciding with the increasing use of these four breeds as outcross sires, scrapie appears to have been declining by 1810 (131), the disease then being considered to be infectious. Little information has been found for the remainder of the 19th century, but the disease has been present in the improved Suffolk breed, derived from the Southdown × Norfolk Horn cross, since the early years of the 20th century, and was a serious problem in many flocks between 1925 and 1950.

Information on the north of England from the Humber and Mersey to the Tweed is very scanty. Comber (82) says that the 'rickets' was unknown in Yorkshire. Bewick (36) makes no mention of any such distemper in Teeswaters, Cheviots and the horned long-woolled hill breeds of the Pennines and the Lake District, among which it has been well known for much of the 20th century, although he was well acquainted with the natural history of the north of England, and especially of Durham and Northumberland.

Scrapie in Britain after 1820.

The years 1820–1850 saw intense interest in the selection and development of British sheep breeds (282, 190), which had become sufficiently 'fixed' for the Royal Agricultural Society of England, which had recognised the Leicesters in the 1840's, to offer separate breed classes for Shropshires, Hampshires and Oxfords from 1859, and for other breeds shortly thereafter (216). Such breed development entailed careful pedigree keeping and selection, which would have revealed undesirable traits such as scrapie. There is no doubt that scrapie had ceased to be of serious concern in southern England by the mid-century, the last reports dating from 1840 (Humfrey (158) from Berkshire; Cleeve (78) from Essex), and veterinary textbooks not mentioning the disease (282). This decline and virtual disappearance of scrapie coincided with the widespread use of breeds considered to be free of the scrapie trait. The Southdowns became predominant in Wessex and in East Anglia (216), while in Dorset, Southdown rams were used on Hampshire-type ewes to fix the new breed of Dorset Downs about 1843. This breed has been and remains free of scrapie in our experience, although one recent case has been reported. Similarly the Oxford Down breed, developed by using Cotswold rams on Hampshire ewes (282), has been and remains free. The Dishley Leicesters were in very widespread use by 1850 all over Britain (216); they had been used extensively by Culley at Fenton, Northumberland from the 1790's, and led to the development in Tweedside and elsewhere in Scotland of the Border Leicester breed, which was recognised by the Highland and Agricultural Society in 1869. The Border Leicester was a recognised cross on Cheviot ewes to give the Scottish Half-bred, which was established in the Tweed valley at least by 1850. As scrapie had appeared in this half-bred by 1853 (see Table 2.1), it may be inferred, on the genetic hypothesis, that the scrapie trait was carried in

both parental breeds before 1850, although no documentary evidence
has come to light; both breeds carry the trait at the present time. The
Border Leicester was also mated with the Scottish Blackface hill breed
to produce the cross-bred type known as 'mules' for lowland commercial
ewe flocks. The term 'mule' was originally restricted to this cross (123)
and is still so used. However, the Border Leicester male × Scottish
Blackface female has for some years become known as a 'greyface',
while the 'mule' is used for crosses of the Blue-faced Leicester male,
derived since 1880 as a distinct breed from the Border Leicester in the
Hexham district of Northumberland, on ewes of the Pennine hill
breeds, the Swaledale, Dalesbred and Rough Fell, and, to a lesser
extent, the Lonk and Derbyshire Gritstone. The distinction is of
considerable epidemiological importance, as the 'mules' bred from
Swaledale and Dalesbred ewes are heavily affected with scrapie.
Although the English Leicesters, direct descendants of the Dishley
Leicester, remain free of scrapie, the scrapie trait has appeared in the
Border Leicester and the Blue-faced Leicester.

In the period 1920–1950 scrapie remained virtually unknown in
Wessex, but it was becoming increasingly prevalent in Suffolk flocks
in many parts of East Anglia, and occasioned correspondence in the
local press at Bury St. Edmunds. The disease had been well recognised
in the 1920's, and by 1950 was a well-known cause of serious loss in
many Suffolk pedigree and commercial flocks throughout Norfolk,
Suffolk, Essex, Cambridgeshire and Hertfordshire. Reliable information for other areas and breeds is hard to come by. A rare published
account is that of Taylor (296) who, with Scottish experience, recognised
the disease in Sussex among Border Leicester × Cheviot commerical
ewes purchased in southern Scotland.

Although the intense interest in livestock improvement from 1780
onwards is as well documented in Scotland as in southern England
(275, 324), M'Gowan (204) failed to find any references to a scrapie-like
illness in his search through many writings of the period 1780–1900; so
we may assume that, if present, the disease was uncommon by English
standards and did not give rise to public concern. Youatt (324) makes
no mention of scrapie, but under the heading of 'scab' he distinguishes
two types, the common form which responds to treatment, and a
second form, less prevalent, which is untreatable. This form may have
been the rubbing type of scrapie. The earliest unequivocal record dates
from 1853 from Roxburghshire and Northumberland and is based on

the recollections, 60 years later, of men told of the occurrences in their youth (63, 286). The details are clearly recalled (286) of an outbreak in Yetholm among a well-known Border Leicester-Cheviot half-bred flock, commencing in 1853 and passing with the sale of breeding stock to other farms, in a manner precisely similar to that encountered today, i.e. associated with the purchase of rams and other breeding stock, some of which became affected and some not. This period coincides with the development of the Border Leicester breed from 1860 (216) and of the South Country Cheviot. No formal records occur until about 1910, although M'Gowan (204) records a report to him by an eye-witness of scrapie in Cheviot sheep about 1874. By 1910 the disease was so prevalent in the Border Counties that a Committee was em-panelled, who in 1912 promoted the investigation by M'Gowan (204) from Edinburgh. At that time the disease was present in Border Leicesters, Cheviots, especially in the Border Leicester-Cheviot (Scottish Half-bred) commercial females, and also in the Scottish Black-face breed (286, 204). M'Fadyean (202) confirms this breed incidence, although it is not clear how much represents his own observations and how much he is relying on Stockman (286). Gaiger (123) from Glasgow gave a comprehensive account of the simple epidemiological aspects, and was obviously well informed on the disease and its occurrence. He confirms its occurrence in the Border Leicester, Cheviot, Scottish Half-bred and Blackface breeds, and also reports its occurrence in Suffolks, Oxford Downs, and Mules (Border Leicester male × Blackface female). He notes its occurrence outside southern Scotland, in Forfar, Aberdeenshire and Caithness, although he notes that in the North Country Cheviot of northern Scotland the disease is rare and very much less prevalent than in the South Country Cheviot, a situation still persisting 50 years later. Greig (135) reported that in 1940 the disease still remained widespread in certain breeds, but was not then seen in the Scottish Blackface.

One may summarise the above by saying that scrapie was probably little known in Scotland before 1850, when it was present in certain flocks of Border Leicester-Cheviot stock on the borders of Roxburgh-shire and Northumberland; but as Stockman (286) reports, strenuous efforts were usually made to conceal the presence of the disease, and many sheepmen did not know the disease when they saw it (204). Under these circumstances knowledge of the disease would only have become public when it had become prevalent and well-established, i.e. about 1900, since when its prevalence has fluctuated over half a century, with absence of clear evidence for its continued presence in the Scottish Blackface, Oxford Down and North Country Cheviot breeds, but its presence being generally conceded in the Border Leicester, South Country Cheviot, and their crosses.

In the 1930's scrapie was again causing increasing concern, and gave rise to the investigations at the Moredun Institute, Edinburgh, where following the demonstration of the presence of an inoculable factor in scrapie by Cuillé and Chelle (87) of Toulouse, the possibility of a contagious agent being responsible was explored (135, 132, 219).

Little practical progress had been achieved in 1950 beyond the evidence already assembled in Germany by May (199), translated and reviewed by M'Gowan (204), and the more precise knowledge of the pathology provided by the excellent study of Bertrand, Carré and Lucam (33) in France, supplemented by that of Brownlee (53) in England. Apart from Stockman (286, 287), who unfortunately did not live to complete his investigations, and Gaiger (123), most writers remained out of touch with the practical aspects of the natural disease. In particular no detailed epidemiological or pedigree studies were undertaken and reported. As experienced sheepmen came to realise that there was little practical gain, and some security risk, the veterinary profession was consulted but rarely. In exceptional conditions, such as the epidemic in the Border Counties of Tweedside about 1900, it was the local agricultural officer who alerted the authorities. As a consequence the veterinary literature on scrapie is meagre, and provides a very imperfect indication of the occurrence, nature and importance of the disease to the British sheep industry in the century 1850–1950, during which the competent accounts provided by well-informed laymen in the previous century were not continued.

Other reports of spontaneous natural scrapie in Great Britain, 1950–1978

Spontaneous field outbreaks of the disease have been reported by Wilson *et al.* (323) and by Stamp *et al.* (283) in Cheviots in Scotland; by Zlotnik and Stamp (340, 341) in a Dorset Down ram in Scotland; by Pattison (236) in Welsh Mountain sheep in South Wales; and by Buntain *et al.* (58) in Swaledales in the mid-Pennines.

These published accounts provide a very fragmented and unrepresentative indication of the extent and nature of scrapie in Great Britain during the period. They describe occurrences of the disease in one or two flocks over a period of a few months. They give no details of any individual identification of affected animals, nor any indication that enquiries regarding family health histories were made, or that such data were available. No long-term epidemiological study has been reported. Studies in laboratory flocks have been made in Scotland (98) on Cheviots and Suffolks, and on Herdwicks in England (Kimberlin, personal communication, 1970).

Scrapie in Germany

The importance of the production of fine wool from the mid-18th century onwards (298) led to the establishment of flocks based on imported fine-woolled Spanish merino sheep by many royal and aristocratic owners in north-central and eastern Germany, in Bavaria and in the Danube Valley. The occurrence of a serious epidemic of *Traberkrankheit* (scrapie) in many of these flocks between 1785 and 1840 in Saxony, Silesia, Bohemia and Prussia gave rise to a substantial literature, much of it recorded in the *Annalen* of the Möglin Agricultural Research Station in Brandenburg, and in Silesian and Bohemian journals (113, 199, 265, 25). The problem was of such moment that a synopsis of 68 pages was printed in Breslau in 1828, derived from items published in the previous 2 years in Silesia, entitled 'A short summary of the trotting disease of sheep as an epizootic hereditary disease' (7).

The central issues were the role of the imported Spanish merinos in the causation of the epidemic; the contribution of various races of merino; and the influence of management practices such as breeding 'in-and-in' and stall-feeding. Some, such as Richthofen (252) considered a skin mite to be responsible, or even the nasal bot (Oestrus ovis larva), but the main emphasis was on breeding practices (199.)

The Saxony State flock of Spanish merinos at Stolpen, near Dresden, was founded in 1765 by a gift of 100 rams and 200 ewes from the King of Spain to the Elector of Saxony, augmented by a purchase in 1778; the original sheep are said to have been very healthy (289). The source *cabañas* are given by Belda and Trujillano (28) on p. 326. George III of England sent a consignment from his English flock to Hanover in 1792. The King of Prussia's merino flock at Frankenfelde in East Prussia was founded in 1815; scrapie was serious in it by 1825 (259). No information has been found on the Hanover flock.

Scrapie seems to have been clearly recognised before 1770 (115). It was often termed simply '*die Seuche*' (the disease or plague) (289). Fink (118), in answering questions from the British Board of Agriculture in London, was obviously well acquainted with the disease over some years in Upper Saxony. It was rampant in Silesia in the 1820's. When a consignment of 100 Silesian Electoral sheep was brought in 1827 to the Royal Farm at Schleissheim in Bavaria, where the existing stock of Negretti and Electoral merinos were at that time unaffected, the disease became rampant for a long period (199).

During the mid-19th century the disease was of very serious concern, as May's account attests. It was most prevalent in Saxony, Moravia and Silesia, i.e. in north-eastern Germany and Poland, where the Electoral merinos were bred for sale in great numbers. Wherever Negretti merinos were kept and few sold, very few cases, or none at all, occurred. May's very detailed and most valuable account of the disease in 1868 underlines its then general prevalence in Germany and Middle Europe, where it was rampant in Silesia and well known in Saxony around Halle. It was still present at the end of the century in Berlin (69); its prevalence has, however, declined during the 20th century. It had become rare in Central Germany around Halle by 1913 (204). Behrens (26, 27) states that the disease was still widely distributed in merino flocks in central and eastern parts of Germany at that time, but that it has disappeared since 1945.

Scrapie in France

No mention of the disease is made by Paulet (239), writing in 1775. Its origin is attributed repeatedly by French writers to the introduction of the Spanish merino studs in the late 18th century. Scrapie is said not to have been encountered prior to their establishment (129), the disease being confined largely to merinos and their crosses. The nature of the serious *vertige* reported in 1788 by Lamerville (178) in Berry remains unexplained. A small flock of Spanish merinos was established at Alfort in 1764, and thence in 1766 under Daubenton at Montbard, where they were crossed with four other sheep stocks, including Tibetan, for which individual pedigrees and all causes of death were recorded (68). No evidence of scrapie had appeared by 1780 (92), and none was seen in the descendants of the Montbard sheep sent to Banks in England in 1785 (68). The main sources of Spanish merinos in

France were the Royal flocks at Rambouillet, which may have been established from the mid-18th century (189), as interest in wool improvement increased (64). The principal foundation stock was a gift in 1786 from Carlos III of Spain to Louis XVI of 318 ewes and nine rams from ten of the best fine-woolled *cabañas* of Spain, the details of which have not been found. In 1797–1800 a large consignment of Spanish merinos, specially selected in Spain by Professor Gisbert of the Alfort Veterinary School, were obtained as official war booty under a secret clause of the Treaty of Basle, in 1797. They were the most desirable sheep from the best *cabañas* (28). These sheep almost certainly came to augment the Rambouillet sheep, which had expanded into ten or more flocks by 1812. *La maladie folle ou convulsive* was recognised at Rambouillet by 1810, although Tessier (297), who was scientific adviser and controller of flocks to the Superintendent of the Royal Domain of Rambouillet, was careful to say he had not observed the disease himself. May (199) states unequivocally that the build-up of scrapie at Rambouillet between 1786 and 1810 was due to 25 years of consanguineous breeding; the definitive contemporary evidence for this view has not been found, but May was in general singularly well informed. Spooner (282) records that the French merinos were not proving as successful as had been expected.

Bernardin, the Director of the Rambouillet stud, writing in 1883 in reply to enquiries from W. G. Markham of Vermont, U.S.A. wrote, as quoted in translation from the 'Vermont Register' by Austin (13): 'In 1840 it was decided to breed for mutton as well as for wool . . . but by 1850 these animals of very large weight were less robust . . . Shortly after 1850 these efforts were renounced . . . and the flock returned to true and ancient type . . . the Negretti type with skin folds . . . Very much folded animals . . . are sometimes weakened in their constitutions . . . As a general rule we avoid (in 1883) giving a ewe a buck of near relation. By near relation I mean the father and his daughter, the mother and her son, the brother and sister. But if exceptional qualities to be perpetuated are found in a male and female of these relations we should not hesitate to couple them if we fail to find in non-relatives the same suitability (*convenance*). For consanguinity is not to be avoided except in cases of individuals having a constitutional vice common to the family.'

From Bernardin's letter we may fairly infer that breeding in-and-in of very close relatives was a well recognised method used at Rambouillet for many years, as May (199) states. Furthermore distinctively different phenotypes had been preferred from time to time, which may well have influenced the pattern of gene frequencies, and certainly the constitution or average life-span, of the Rambouillet sheep. When Markham visited Rambouillet in the early 1880's he found the sheep to be of American Delaine type, relatively free of skin folds. The sheep were tattooed in the ears, and pedigrees kept; any families of weak constitution would have been readily identified and culled.

The stud at Malmaison, formed by the Empress Eugénie, was based on Negretti and Paular blood lines and was free of scrapie, as was a flock of similar breeds in the Landes, and it is said to have remained free (199). Whether the later outbreaks in milk sheep at Alfort in 1819 (129), in Seine-et-Oise (31) and in Larzac milk sheep in southern France (253) occurred in animals bred with an infusion of Spanish merino blood, or in contact with merinos, is not recorded. The impression is that an exotic pathogen was responsible, but the evidence is inconclusive, and the original presence of the scrapie trait in the indigenous breeds cannot be excluded. The report of Cauvet (70) probably refers to another disease.

By 1870 the main features of the disease in French sheep were well recognised (30). Outbreaks were often serious economically (34), and sporadic outbreaks occurred from time to time until 1950 (67, 61, 33). Almost all these outbreaks occurred south of the Massif Central (263). The Berrichon breed (probably the du Cher type) and the Lacaune were involved, but in general the breeds are rarely stated, the implication being that it is a curious infectious disease. Trasbot (304) reported an epidemic in the Sologne in 1889 in a flock of Southdown *solognots*—cross-breds based on mixed breeds, described by Malingié-Nouel (192) as of little intrinsic merit. It is difficult to know how much blame to attach to the Southdowns in this outbreak, or in that recorded in the Silesian newspaper report of 1864 (199) since we do not know the local breeding patterns. Lucam *et al.* (191), reporting on cases in 1949 in a flock near Chateauroux in L'Indre, in Berrichons with some admixture of Southdown blood, consider that the disease is and has been for many years more widespread and prevalent north of the Massif Central than is generally admitted. The flockowner recalls the disease in Berrichons in 1919–1921. This report is also of particular interest as the flock was of two distinct provenances. One part, which had been on the farm for 100 years, was unaffected, and it was in the other part, purchased in 1947, that the cases occurred in 1949.

The thesis of Saurat (263) from Toulouse provides a most excellent and revealing summary of the literature and of his own studies on *la tremblante,* together with additional data on the transmission experiments of Chelle and his colleagues in the late 1930's. The work is little known, and rarely quoted, doubtless because of the exigencies of war at the time of publication.

Scrapie in Spain

It seems probable that the distemper recorded as widespread in migratory sheep in Spain, and called *das Drehen oder Traben der Schafe* by Stumpf in 1785 (289) (rendered as 'dizziness' in the English translation) or *vertige* (299) was in fact scrapie. Although no unequivocal account is said to occur in the Spanish literature (27a), there are good grounds for the suspicion that outbreaks or occurrences may have passed unrecorded. A form of 'scab' unresponsive to the usual treatment occurred in merinos exported to the United States of America in 1812–1814 (68), and was associated with cachexia and a high death rate; it may have been scrapie. Youatt (324) also records a similar form of 'scab', resistant to treatment, and thought to be scrapie, in Scotland. Its occurrence in the merinos exported to Germany and their descendants suggests that certain *cabañas* (such as the Escorial) originating from the Central Madrid Uplands were more likely to carry the trait than sheep from the Negretti (Infantado) and Paular *cabañas* of the Spanish west and north-west (199). The transhumate (itinerant) fine-woolled merino *cabañas* (46, 173) were affected, but not the non-travelling (*estantes*) common sheep stocks (289). Recent data are scanty, but the disease is said not to be present now, or to have been recorded previously, in Spain, possibly because of the social and geographic isolation of their flocks (28, 46). It is of particular interest that Stumpf (289) comments that 'dizziness' is reported only in these special merino 'travelling' flocks, and never in the common settled native sheep of Spain; the evidence for this distinction, and a more detailed account of the malady, would have been welcome.

Scrapie in Hungary and the Danube Valley

In Hungary the disorder was unrecorded until the early 19th century (137), when a series of revealing papers appeared. In 1843 an anonymous writer on sheep breeding (9) gave a short and very illuminating account of the trotting or 'wobbly' illness with wasting, nibbling and rubbing, and clearly distinguished the latter from the *rágá* or biting illness (rabies). The disease occurred in 2–3-year-old sheep in certain flocks, and was inherited; 'do not buy a ram from such a flock'. Prély (245) confirmed these views, and the hereditary basis, and added that although it was commoner in merinos it was 'an ancient illness present before the merinos came' and not imported with them. He emphasised the early inconspicuous stage of being easily startled or frightened, and of 'cowardly' behaviour. With no cure, affected animals should be slaughtered. Zlamál (330) of the University of Pest agreed, and stressed the scraping along the whole length of the animal (*Wetz-krankheit*). All affected animals *and all their relatives* should be destroyed, i.e. the heredo-familial nature of the complaint was clearly recognised. Also from the University Veterinary School, Eisenmáyer and Göbel (109) reinforce these observations, drawing attention to the three principal types of the disease, the scraping or brushing (*Wetzkrankheit*), the startled–stupid (*Schrecksucht*) and the trotting–wobbly (*Traber-krankheit*). They record its occurrence in goats as well as sheep. Proneness in sheep is related to over-breeding of 'too-refined' types, and occurs especially in strains capable of very good food conversion, and under suitable feeding of developing very rapidly, i.e. high live weight gain for age; it is accentuated by over-heating of pens. Zlamál (331) reaffirms the views of the previous writers but adds that it was the Electoral merinos of fine-woolled strains which became affected, the disease being unknown in ordinary coarse-woolled home-bred sheep, i.e. those without any strong infusion of fine-woolled merino blood. Progeny of affected ewes are frequently affected, but the dams of affected offspring are frequently not. Autopsy examination reveals little of significance. Hutyra (160) gives three lines only to the topic, indicating perhaps its diminishing importance. Indeed the disease seems to have become so uncommon in the present century that Aldásy and Süveges (4) reporting a serious outbreak observed in 1959, in a large flock from which breeding stock had been introduced to a cooperative farm, claimed that theirs was the first account of scrapie in Hungarian sheep; in both flocks losses of 6–25% occurred, and both flocks were slaughtered. The origin of this outbreak could not be traced and, except for this occurrence, the disease has not been reported in Hungary for at least 40 years (293). Karasszon (169) in an historical summary traces the origin of the disease in Hungary to

Escorial merinos sent as a gift from Charles III of Spain in 1748, and provides a number of additional references. In Austria the disease was observed in the late 18th and 19th centuries (254), but it appears not to have been so severe as in Silesia and Saxony. In the eastern Danube no information has been found for Romania, Yugoslavia or Bulgaria, apart from a single outbreak in Bulgaria among imported sheep in the 1960's (161). The indigenous breeds include some merinos and their crosses (142), which are considered to be free, but no evidence for this is adduced.

Scrapie elsewhere in Europe

Scandinavia

The disease has not been recorded in indigenous sheep in Norway (207, 209, 217), Sweden (276, 242), Denmark (318) or Finland (200); but two imported Suffolk rams developed the disease in Norway in 1958–1959 (208, 209) and possibly an imported Cheviot ram in 1940 (217).

The Netherlands and Belgium

In the Low Countries (the Netherlands and Belgium) sheep were of small account, and no mention is made of any disease problem (97) until the 1960's, when scrapie was diagnosed in the Netherlands in an indigenous sheep of the Texel breed (314). These authors concluded that the disease must be considered endemic in certain Dutch sheep, with annual attack rates of 1–4% in certain flocks of Texels. In Belgium scrapie has been recognised in an imported Suffolk ram, and later in three of his pure-bred daughters (153).

Eastern and southern Europe

An outbreak in an East German pedigree merino flock was reported in 1973 (147). No information is available for Russia and Poland, nor for Italy, Portugal and Greece.

Iceland

In Iceland, with the specialised northern type breeds related to the Herdwicks of Britain and the Tauter of Norway, no scrapie (rida) was recorded until the 1940's, when the disease appeared in sheep associated with animals (breeds not known to me) imported from Germany in 1934 (222, 221).

II. Scrapie in Sheep Populations outside Europe

Information on the indigenous breeds of the main sheep-breeding countries prior to the European expansion of the 16–19th centuries, namely from North Africa (112), the Middle East, India and Central Asia, has not been found. The only account of scrapie in an Asian sheep stock is from India, following the introduction of Rambouillet rams from France in the 1920's (337, 170, 170a). An important development was the settlement of the Savanna grassland regions of the subtropical and temperate zones of the world by sheep imported from western Europe, notably from Britain, Spain, Germany and France. Sheep were exported in considerable numbers after 1808, following the disruption of the Spanish merino flocks in the Peninsular War by the French army's advance across Central Spain in 1808–1809, and in the period of expansion after 1815.

Australia's national merino flock of 300 million head is derived from sheep from the Cape of Good Hope, and from George III's merino flock in England, through the MacArthur importations to their flock at Camden, New South Wales, in 1808–1815, and later importations from England, France and Germany (13, 68). Many specifically English breeds have also been imported. No case of scrapie was known up to 1950, or in the merinos at any time; but an outbreak occurred among imported British Suffolk sheep in 1952 (56). In New Zealand, with a national flock of 25 million head, largely based on English Romneys, Lincolns and Southdowns, with merinos from Victoria in the 1850's, no case had been recognised by 1950 (55). An outbreak, again in imported Suffolks, was reported by Brash (44, 45).

In the United States and Canada, populated by many British, German and French breeds, the disease was exceedingly rare up to 1950, although there are suggestions that an occasional un-reported case may have occurred. For more recent outbreaks, see Stuart et al. (288), Wagner et al. (317) and Hourrigan et al. (156). In the Argentine, the national flock contains many Romneys and Romney crosses, but importations of merinos, e.g. from France in 1890 (13) and Australian breeds have been made. The disease is unknown (62). In the Falkland Islands, with their important sheep stocks based on northern Scottish Cheviots and Romneys from Britain and Polworths and Corriedales from Australasia and South America, no case has been recognised (320).

In South Africa, which received merinos from many sources from 1790 onwards (180), the disease has not been recognised in merinos or in indigenous African breeds (112), and there was no evidence of the presence of scrapie before 1950. An outbreak in imported Hampshire Down sheep was reported by van der Merwe (315). The same breed was involved in an outbreak in Kenya in 1970 (83). For Mediterranean Africa, which probably provided the foundation Berber stock for the Spanish merinos about 1146 (173), no information has been available.

In summary, reports of natural scrapie in sheep have appeared, mostly since 1950, in many parts of the world (Table 3.2). The focus of endemic disease in North-Western Europe has continued, with wide fluctuations in prevalence, often poorly reflected by the published records. Of particular interest is the occurrence of the disease in countries in which the disease had not previously been recorded, often associated with the introduction of pure-bred breeding stock from

Table 3.2. *Recorded occurrences of natural scrapie outside Europe, 1930–1980*

Region		Date	Reference	Breeds affected
North America				
Canada	Ontario	1938	263a	Suffolk, imported from U.K.
U.S.A.	Michigan	1947	302a	Suffolk, from U.K. and Canada
	California	1952	288	Suffolk, imported
	Ohio	1952	317	Suffolk, imported from Canada
	Many states	1960	156	Mainly Suffolk: also Cheviot, Hampshire and Montadale
South America				
Colombia		1968–1971	see p. 153	Hampshire and Dorset Down, imported from U.K.
Brazil		1977		Hampshire Down (same flock of origin as in Colombia)
South Africa		1964–1972	315	Hampshire Down, imported from U.K.
East Africa				
Kenya		1970	83	Hampshire Down, imported from U.K.
Australia		1952	56	Suffolk, imported from U.K.
New Zealand		1952–1954	44, 45	Suffolk, imported from U.K.
India				
Himalayan foothills		c. 1940	337, 170a	Local mountain breed, with imported Rambouillet stock

countries with the disease, and indeed from breeds and source flocks now known to have been heavily affected at the time the emigrations took place.

III. Some Epidemiological Features of Scrapie in European Sheep, 1700–1950

Although the data are fragmentary when viewed on any national flock basis, they may be used to provide a rough indication of the occurrence of scrapie disease in certain breeding populations, and by inference of the prevalence of any traits influencing the appearance of the disease.

In such an analysis the principal points of interest are (1) the wave of high prevalence between 1750 and 1820 in England, France and Germany; (2) the slow decline between 1825 and 1875 in France and Germany, and the rapid decline and virtual disappearance in Southern England, but with the recognition of the disease in Scotland; (3) a continuation of the decline in Germany between 1875 and 1925, the continuing sporadic occurrence in France, and an increase in Scotland and, at the end of the period, in England; (4) since 1925 the virtual disappearance of the disease in Germany, its continuing fluctuations in France, and a patchy localised increase in parts of England and Scotland. During this period the disease was virtually absent in the Southern hemisphere and in other regions populated by the descendants of the many importations of European sheep.

The wave of prevalence in Germany, France and England between 1770 and 1820 may seem at first sight to suggest a long and slowly unfolding infectious epidemic, commencing in England and spreading to Germany and thence to France. However, closer study reveals that the factors underlying the outbreaks in the three countries differ in important respects, and that general matters of breeding policies, sire selection, on-farm management practices, economic goals and improvement aims, be they for wool or for meat, and pressures to multiply scarce and desirable breeding stocks, are of critical importance.

In Germany scrapie was already increasing by 1760, i.e. before the first main importation by the Elector of Saxony of Spanish merinos to Saxony in 1765 (118). Although small numbers of Spanish merinos may have reached Southern Germany through Austria via Leghorn and Trieste before the Elector of Saxony's importation of 1765, it is probable that the indigenous native German sheep stocks already carried the scrapie trait (181, 115). Thus the German epidemic is unlikely to have been solely due to an exotic infection introduced with the Spanish merinos (130, 104, 26, 27).

May (199) records a much greater prevalence in merino sheep of the light-boned Electoral strain in eastern Germany and Poland, where the heavier Negretti sheep are said not to have been affected. It may be noted that in the Prussian State Flock at Frankenfelde the sheep (of Electoral type) were housed for much of the year. In southern Germany, however, neither the Electoral nor the Negretti merinos were affected, whether in the mountains or in the low ground, although a flock of Silesian merinos, introduced into a Bavarian flock of mixed Electoral and Negretti crosses, continued to be affected. There are numerous statements that the disease is hereditary (118, 259, 7, 199), although some rams producing many affected progeny are not themselves affected. Fink (118) records the successful eradication of scrapie from his own flock in Saxony by the selection and use of sound and healthy rams from flocks perfectly free of the disorder. For many decades subsequently the informed opinion was that animals suspected of the disease must not be used for breeding, and that they, and any progeny of animals which become suspect, must be killed at once 'sparing them not'. A similar policy was advocated by Zlamál (331) in Hungary. Any rams or ewes must only be bought-in from unaffected flocks, and that status accepted 'only after the most careful investigation, since every effort is always made to conceal the presence of the disease'.

The prevailing view at the time was of a disorder occurring mainly in certain breeds, especially merinos, but only in certain strains; that epidemic outbreaks tend to be associated with close inbreeding practices carried on for one or two decades previously; and that environmental factors are not obviously involved. The experiences of Fink (118) at Cositz in Saxony, of Stumpf (289) in Anhalt-Dessau, and of Thaer (299, 300) in Saxony and Prussia provided evidence of the practical control of the disease by careful ram selection and the avoidance of inbreeding. In France, as in Germany, the greatly increased prevalence of *la tremblante* or *le vertige* after 1800 followed the introduction of the Spanish merino and was especially associated with the Rambouillet flock, derived from ten élite Spanish *cabañas*, while the Malmaison flock, based on the Negretti and Paular *cabañas*, remained unaffected (199), as did the earlier Montbard flock. As to the nature of the condition, Girard, the director of the royal veterinary school at Alfort, has this to say:

> The theory of the heritability of the disease seems to us now well-established, and confirmed by experience. We cannot accept the opinion of those who claim that the condition is capable of being transmitted by contagion, or that a ram may contract the disease by tupping an affected ewe (129).

He proceeds to recommend practical measures—early culling of affected individuals, buying in rams from healthy flocks, etc.—for eliminating the disease, which is clearly recognised as incurable.

IV. The Role of the Spanish Transhumate Merino in the International Dissemination of Scrapie, 1750–1850

The available evidence raises serious doubts regarding the assertion, often made, that there were sudden and discrete introductions of Spanish merinos into north-west Europe between 1750 and 1800, and that these Spanish sheep were responsible for introducing the disease, which had previously been unknown in the indigenous unimproved sheep. Some merinos may well have carried the scrapie trait (alternatively, the infective agent), but equally clearly others, e.g. those coming to Scandinavia, Britain, Bavaria and Württemberg, did not. The real question is why certain merino populations associated with the stud flocks at Anhalt-Dessau, and at Stolpen in Saxony, Frankenfelde in Brandenburg, Branzdorf in Silesia and Rambouillet in France, became very severely affected after a period of about 20 years (199), while other merino stocks in Germany, France and Hungary remained completely free, as far as the records indicate.

The answer seems to be that the breed of Spanish merino was crucial to the appearance or non-appearance of scrapie. In brief, the Negretti and Paular merinos were not affected, while certain Escorial merinos, and their Electoral descendants which came to Germany and France, were affected. No unequivocal evidence of dissemination before the F_2 generation of a cross-breeding of Electoral grandparents from affected stock (e.g. Silesian sheep introduced into the Schleissheim Negretti flock in 1827) has been found.

If this information, incomplete as it must be, is considered in terms of a recessive gene-determined disorder, it is at least compatible with the assumption that the Negretti and Paular *cabañas* had a very low or nil scrapie-allele frequency, and as these *cabañas* provided many of the original foundation merino stock of Australasia, America and South Africa, one may expect the disease to be absent or very rare in their descendants, as indeed is the case. It appears that the later introductions of French Rambouillet stock in the 1850's and 1860's (13), and of American Vermont merinos between 1880 and 1910 (274) into Tasmania have not significantly affected the frequency of the disease, for it has never been reported in the Australian merino.

This interpretation is in accordance with the evidence that *Traber-krankheit* was present in Germany before the main introduction of merinos, and need not be attributed to an exotic infection introduced with them. The weight of the German evidence (215, 130, 104, 26, 27), favours the view that scrapie was already present in the indigenous sheep of north-east Europe and the Austro-Hungarian Empire before the importation of merinos from Spain. Some merino *cabañas*, notably of the Escorial breed, appear to have carried the scrapie trait; but *Traberkrankheit* did not occur in Germany for several generations of inbreeding. Cross-breeding of merinos with indigenous Hungarian sheep over a period of 50 years provided no reports of the disease (137); when it appeared (9), the outbreak was severe, and persisted for nearly 50 years (245, 330, 109, 160). The disease then disappeared (293) and has not recurred, apart from a single outbreak in 1959 (4), in sheep recently introduced from England. The English records of 1750–1850 establish the importance of the breed structure of the populations in determining the incidence of scrapie, and the extremely diverse contributions to the overall national or flock scrapie attack-rate made by the animal subpopulations of different breed constitution, some breeds and their first crosses being heavily affected and some apparently free of the complaint. Assertions are repeatedly made that the ravages of the disease may be mitigated by eschewing consanguineous matings, by changing the sires frequently, and by careful selection, after meticulous enquiries, of sires from new blood lines in other flocks free of the scrapie trait, and as a last resort by replacing the flock with stock free of the trait, often of another breed known to be unaffected, or by using rams of an unaffected breed as a top cross and retaining their F_1 progeny as flock replacements.

The German and French experience confirms many of these views. The high attack-rates which developed over 25 years in the Stolpen and Rambouillet merino studs were attributed by May (199) to close inbreeding and repeated consanguineous matings of animals from the same flock, a practice dictated by the desire to fix the type of sheep and to multiply numbers as rapidly as possible, particularly in periods of high market demand, such as existed for the Electoral sheep in Eastern Germany and Central Europe and in France around 1800. Both State flocks were expected to provide rams for use in general agriculture; for example, in 1811 the ten French Imperial merino stud flocks, which included Rambouillet and Malmaison, with 2600 breeding ewes, were producing 1200 rams for distribution (68), a figure which allows but little scope for culling undesirable traits, since Spanish merinos usually had only a single lamb (289, 118). A similar breeding policy was also adopted at Frankenfelde in Prussia from 1815. Scrapie was rife within a decade—a time interval for the development of a serious incidence of scrapie also observed in English flocks since 1950. These accounts also speak of 'hereditary trotters', which are not obviously affected themselves but which may leave many progeny affected (199). The progress of the disease in a flock is slow and irregular, but once it occurs in the animals kept for breeding, the disease continues and rarely dies out, but spreads over a decade or longer, until the young stock fail to make up for those lost by death (199).

Both French and German observers suggest that frequent and excessive sexual activity by the rams predisposes to the disease. This however is probably an inference from the heightened libido exhibited by rams in the preclinical and early clinical stages of the disease (see below, p. 64), and is an early consequence of the malfunction of the hypothalamo-neurohypophysial system (see pp. 121ff) rather than a factor in the causation of the disease. Geographical location and altitude were regarded as unimportant. The effect of the close housing, with feeding on chopped feed, practised in East Prussia and Silesia for 6–8 months of the year and in some cases continuously (e.g. at Frankenfelde) on merinos imported from Spain might have been an important factor. In Spain they were not housed, and over 500 years under the control of the Mesta had been accustomed to a regular movement along *cañadas* (stock routes) of natural semi-arid grazings in the open, limited amounts of salt being available over the three late summer months only. Yet the Saxony merinos, housed similarly, remained healthy and vigorous (118).

Of considerable interest are the different prevalence rates of scrapie in the merino stocks derived from different merino *cabañas*, which might number 10,000 to 44,000, in the Spain of 1780 (see p. 27). The Negretti and Escorial *cabañas* were each of about 40,000 head. Compared with many European sub-types, e.g. the short-woolled Down breeds today, each *cabaña* may be considered a distinct breed of merino (as indeed they were on many phenotypic characters) and it is probable that rams were rarely if ever selected from outside a *cabaña*, since 30 or 40 separate strains from separate *dehesedas* would be available for suitable out-crossing, in the manner practised between the individual British pedigree flocks today. Thus one may reasonably consider the Negretti, with its heavier body and bones, as a breed distinct from the light-boned Electoral, and infer that sheep imported from these *cabañas* would be of distinct strains, depending on the *dehesedas* from which they came. Such a view may offer an explanation, in terms of the English breeding experience, of why the Negrettis and Paulars were said to be free of scrapie, while some Electoral strains were heavily affected, such as those in Saxony, Prussia and Silesia, while others in Bavaria were not affected. In short, of the different breeds of the merino race, some were affected and some not, and within an affected breed some strains or blood lines were affected and some not, as among English sheep. That scrapie was not a general environmental hazard affecting all sheep in Spain is shown by the absence of scrapie in the common non-migratory *churro* sheep (289).

This variable prevalence of scrapie in various merino stocks may hold the key to the generally accepted view that scrapie has never been observed in merinos in Australia or New Zealand or South Africa, and has certainly not been prevalent within living memory, although Macarthur's importations of 1808–1812 came from an area of England in which scrapie was prevalent. The merinos which formed the foundation stock in the Antipodes and southern Africa were mainly of Negretti and Paular stock, with probably a very low prevalence of the scrapie trait. The merinos which went to the United States in 1809–1812 were also mainly of Negretti and Paular stock, with some Gaudaloupes (68).

V. Contemporary Views

The balance of opinion on the incidence of scrapie in Germany and Central Europe over the period 1750–1820 may be summarised as follows:

(1) The disease was probably present before 1750 in the indigenous sheep, i.e. in flocks unimproved by the introduction of merino blood.
(2) The introduction of Spanish transhumate merinos after 1760 was associated with a marked increase of the disease, especially in the descendants of the finest-woolled strains of Escorial and Electoral breeds, but not in the Negretti breed.
(3) The disease was inherited, and knowledgeable breeders recommended the very careful selection of rams from flocks free of the disease.

The practice of close inbreeding with rapid population increase, widely used between 1765 and 1820 to make the most rapid use of the valuable imported Spanish sheep, was considered a very important factor in the greatly increased prevalence of scrapie.

Three principal means of mitigating the ravages of the disease were repeatedly recommended, and are given below.

(1) If your existing flock is to be maintained as a pure breed, take exceptional care in selecting your rams from flocks free of the 'distemper', and if necessary replace the entire flock from a source flock free of the disease every 8–10 years (10, 308), change rams frequently (313) and select the sires in such a way as to avoid close inbreeding (95); and do not confine the selection to your own flock more than is absolutely necessary (129).
(2) Change your flock from the affected pure breed by outcrossing with rams of one of the unaffected improved breeds, and thus gradually change the breed composition of the flock towards the new improved breed (308, 41, 95).
(3) Avoid the retention of the progeny of affected animals or of rams unaffected but producing many affected progeny (118, 109).

VI. The Dispersal of Scrapie from Western Europe

The most remarkable aspect of the historical records is the virtual absence of evidence for the dissemination of scrapie until the 20th century. Although many of the temperate regions of North and South America, Australia, New Zealand and South Africa became populated with sheep imported from Europe, notably from Britain, Germany, Spain and France, between 1800 and 1850, records of any rubbing disease, other than scab, are hard to find (see pp. 50–51). Yet by the middle of this century the disease had appeared in North America and other parts of the world, principally in Suffolks and Hampshire Downs recently imported from Britain. This pattern of occurrence or non-occurrence of scrapie is difficult to explain on any simple hypothesis, genetic or other. Why did not scrapie appear among the European sheep exported between 1790 and 1820, a period when in much of southern England the disorder was widespread and endemic, as it was in certain merino stocks in Germany and France, at the very time when Australia and South Africa and the United States were being stocked with their merinos as well as with other breeds? Similarly, when the Cheviot exportations of the late 19th century were made, the disorder was already widespread in many parts of Scotland, yet the disease did not occur in the New World for many years. And why have the Romneys and Southdowns and Suffolks of New Zealand and of the Argentine failed to develop the disease? The scrapie trait, whether infective or genetic, has behaved in its present capricious, variable and unpredictable manner for a very long time with apparently little change. It is the understanding of this often bewildering behaviour which is central to the scrapie dispute, and essential for effective control and prevention of the disease.

Chapter 4

Clinical Features of Natural Scrapie

The 18th and 19th century writers provided general accounts of the disease, which are clear and readily recognisable, and were well reviewed by May (199). Stockman (286) provided a more detailed survey, based on enquiries and on his examination of about 30 affected animals, since when no comprehensive study of the clinical manifestations of scrapie in sheep has been published. In 1950 I initiated some enquiries, which indicated that the forms of the disease were more diverse than the veterinary literature had led me to expect, and that affected animals often exhibited a variety of syndromes, which are summarised on pages 2–4. The following account is based on personal observation of about 2100 affected animals, mostly from flocks under long-term supervision, in which the normal behaviour of unaffected sheep was well established, supplemented by close observation in the hospital paddocks in full view of my residence. Cases from the following affected breeds and half-breeds have been examined clinically, and representative examples pathologically: Suffolk, Border Leicester × Cheviot, Kerry Hill, Clun Forest, Hampshire Down, Teeswater, Dorset Horn, Swaledale, Dalesbred, Welsh Mountain, Herdwick, and Masham Mules. Normal controls from unaffected breeds include the Southdown, Romney Marsh and Dorset Down.

I. Premonitory Signs and the Preclinical Stage

The onset is usually insidious, and it may be several months before definite illness can be recognised. During this premonitory stage one may observe changes of behaviour, lasting for short periods at irregular intervals. These are likely to pass unrecognised. A sheep will stand quietly, head raised in a fixed position as if looking ahead at some object, but with eyes immobile and unresponsive, in a fixed stare. An animal may be unduly restless; in a flock grazing quietly on a good grass sward the animal takes a few mouthfuls only, and moves on to a new grazing site; or at a well-filled trough at a fixed daily feeding-time it is continually moving, unlike its fellows, to a new feeding position; or when the flock is resting and ruminating it will get up and change its position; or when a resting flock is made to rise, it will move rapidly amongst its fellows and turn to face the intruder, as if unduly apprehensive or startled. Some animals become difficult to herd and drive, and unresponsive to the shepherd's dog, and may charge the dog or a gate; they are restless when restrained, and cases are frequently first noted during shearing by their small 'jittery' movements, which are unusually troublesome to the shearer.

In closely shepherded flocks experienced shepherds recognise the possible implication of such behaviour, and cull the sheep before definite signs are evident, and thereby may keep their flock 'free from scrapie', i.e. free from clinical evidence of the disease.

II. Clinical Stages

The Early Clinical Stage (Stage 1)

The early signs, though diagnostic to an experienced close observer, are frequently inconspicuous; on casual inspection of the flock the sheep looks perfectly normal. The fleece may show a whitish tip to the staple over the back, and may become more open, with loss of springiness and of 'character', and with a harsh 'handle', indicating some subtle metabolic and/or nutritional alterations in the skin.

Exercise tolerance is reduced. Affected animals become fatigued when driven quickly, and may collapse in their tracks as they walk, recovering after a short rest. Later the gait becomes clumsy and the animal is slightly unsteady when forced to turn sharply off its hindlimbs. At this stage neurological examination reveals no positive evidence of abnormality. Loss of bodily condition, as determined by visual inspection and digital palpation, is usually not present, but some disturbance of water metabolism is common, with the animal going to water frequently but only ingesting small amounts; when present, this behaviour is highly suspicious. After 3 to 4 weeks, i.e. towards the end of stage 1, the sheep frequently begins to rub itself against hard firm objects, the common sites of rubbing being the poll of the head between the ears, and the buttocks on each side of the base of the tail, rostral to the ischial protuberance and lateral to the dorsal processes of the vertebrae; or it may occasionally scratch with its hind feet just in front of the shoulder-joint or behind the elbow.

The Middle Clinical Stage (Stage 2)

Some 2 months from the onset of stage 1, frank signs of the disease are presenting. The animal is 'unthrifty' i.e. is not carrying so much flesh as normal sheep in a similar environment and with similar access to food; and its fleece and hair look dull and lustreless. There may even be some wasting of the skeletal muscles. The animal fatigues quickly when driven, and 500 yards is about its limit for slow walking. Its gait is ataxic, with excessive flexion of the fore and hind limbs, which tend to be kept as far apart as possible, giving the animal a wide-based stance. The head is held up, with the nose in the air. The ataxia is accentuated if the sheep is hustled, and it becomes excited and upset much more readily than a normal animal. The signs of rubbing are clear-cut; the original wool over the rubbed parts becomes matted, with loss of the outer part of staple and even shedding of the wool fibres, which are replaced by a new short growth, usually pigmented in dark-faced breeds (see frontispiece). The rubbing of the woolled body is confined to localised sites round the tail base, the lateral thorax just above and behind the elbow, and the lower lateral neck. Occasionally a general brushing of the side of the whole body is the principal sign (the 'shrugginess' of older writers). An affected animal will walk considerable distances to indulge in bouts of rubbing lasting 2–5 minutes. The sheep nibble at the haired skin below the knee and hock, scratch like a dog with fleas, and often rub the bridge of the nose.

The 'nibbling' response can be elicited by firm digital pressure, sufficient to involve the underlying skeletal muscle, applied backwards and forwards along both sides of the dorsal processes of the lumbar spine, the lumbosacral region of the base of the tail and the buttocks over the ischial protuberances. This response consists in slight raising of the head and nose (with obvious concentration of the animal's attention), nibbling movements of the lips with extrusion of the tongue, and smacking of the lips—the whole attitude suggesting considerable satisfaction to the sheep. It is difficult to elicit this response by light touch of the skin over rubbed areas, but it may be readily elicited by deep muscle pressure, and is abolished by the infiltration of a local anaesthetic agent or epidural spinal anaesthesia. Over the haired parts of the skin a rash frequently appears, with papular eruptions. The papules become capped with a serous exudate, which hardens and is shed, leaving a small firm pimple, from the surface of which the hair is shed and does not regrow. The appetite is still good, and a limited series of observations on the individual intakes of food indicates that these sheep probably eat enough food to keep them in good health if they were otherwise normal; detailed metabolic studies have not been made. Affected animals often drink frequently, in small amounts, and probably in excessive total quantities. In one series of individual observations normal sheep of one breed housed on dry feed in mid-winter were drinking about 5–6 litres of water per 24 hours, while affected neighbours drank 10–12 litres, and some drank a 5% solution of sodium chloride in preference to the water.

The postural and tendon reflexes are present. Response to pinprick of the integument is normal: sometimes there seem to be some ill-defined areas of hypoaesthesia, but no constant disturbance of cutaneous sensation has been found. The ataxia is little altered by blindfolding. Defaecation and micturition are normal; the eye-preservation and pupillar light-reflexes, hearing, and deglutition are normal. The voice is usually normal, but in some animals the bleat has a special high-pitched timbre.

The Late Clinical Stage (Stage 3)

By 3 or 4 months after the development of the first signs, the florid stage of the disease develops, which ends fatally in from 2 to 4 weeks. The sheep waste away until they appear to have very little flesh, and the muscles are small and flabby. The sheep lie about a great deal, and cannot walk more than 50 yards at a time. They are now extremely difficult to handle because they cannot be driven like normal sheep, and they become very confused by any change in their routine of living. For example, they become very agitated if they are separated from known companions and moved into another pen with different companions, even within the same small building; such a change may frequently lead to their death overnight.

Their gait is extremely unsteady, and they fall over if they attempt to turn sharply. Muscle power is not lost; they press firmly against human restraint and can jump; for example, an affected ram was seen to clear a 4 ft 6 in (135 cm) high oak farm gate 4 days before its death. Frequently they try to advance at a sort of clumsy gallop, moving the two hind limbs together like a springhorn antelope, because they lose their balance with only one leg on the ground. The patellar, Achilles, triceps, biceps, and plantar reflexes of the limb muscles are usually present, though they may be weak, but the muscles on which these reflexes depend are seldom severely affected. At rest the affected sheep adopt normal body postures and, after a fall, they regain their usual posture, although in a clumsy manner.

In the terminal stage the sheep cannot rise to their feet, but if housed indoors they may live for upwards of a week lying on their sides. They now often show hypertonicity of the limbs, particularly of the side on which they are lying, which become very difficult to flex. Such recumbent animals may eat if food is brought to them, but they are usually incapable of doing so. They can swallow if fed artificially. The invariable end of this stage is death; in the field, animals are merely found dead after a cold wet night. Death usually occurs within 4–6 months of the onset of clinical signs. Occasionally the disease may progress to death within 2–3 weeks, if the onset coincides with the end of pregnancy, while rarely the disease may last for 12 months and even longer.

The appetite is usually good until the end, but an occasional sheep appears to lose its appetite in the last stages. The stools appear normal.

The disease does not seem to affect reproduction in the female until the loss of muscle power seriously interferes with her activities. Normal lambs are carried to term, parturition is normal, and lactation is usually uneventful, with a good milk-flow as judged by the growth of the lambs. Occasionally, when parturition coincides with an early stage of the clinical disease, the mothering behaviour is lost: the ewe becomes confused and refuses to have anything to do with her lambs. Males are active and fully fertile 2 months after the onset of the clinical illness; indeed their libido is maintained fully to the terminal illness, and probably their fertility also. In general, they are more active at mating time than many rams of the same stock.

Simple non-instrumented studies of affected and unaffected sheep in pens have shown that the ability to maintain body homoeostasis may be severely impaired in affected sheep. The diurnal variation of body temperature, which is normally about $38 \pm 1°C$, is wider, and short periods occur when the rectal temperature is $40°C$. Affected sheep also show greater and more easily provoked rises of heart rate, often to frequencies of 140–150 per minute, when the controls rise to 100; i.e. the resting rate of about 80 per minute is little affected, but tachycardia supervenes more readily, and with frequent arrhythmias.

III. Other General but Less Common Clinical Features

Certain variations of the clinical pattern are not uncommon. Thus there may be different presenting signs. A sheep may suddenly refuse to be driven or respond to a sheepdog, as if temporarily confused and aggressive; modest but definite swellings of the muzzle and face appear and then subside; sudden swellings of the ears develop, due to haematomata, which heal, leaving a typical distorted 'cauliflower' ear; in many cases the ears flap unduly when the sheep is walking; the eye movements are jerky, and the animals have an unusual 'fixed' stare (in these, severe myopathy of the extrinsic eye muscles has been found at necropsy). In a few cases the gait has been stiff and hypertonic, resembling that seen in myotonic dystrophy in man. In the later stages the soft tissues of the bridge of the nose are not infrequently swollen; occasionally signs of disturbance of alimentary motility occur, with recurrent ruminal tympany or frequent diarrhoea, not shown by normal animals, and sufficient to lead to unexpected death. One group of six 'affected' shearling females all developed a fatal lungworm (*Dictyocaulus*) infestation, while the remainder of the age group, housed together in the same way, remained clinically unaffected. However, convincing evidence for any special liability to helminthiasis in scrapie-prone animals has not been found. A few sheep show weakness of the dorsal neck muscles, so that they rest the lower jaw on the ground, and cannot raise their heads off the ground for more than 2 or 3 minutes; marked pallor of the dorsal anterior neck muscles is commonly present in these animals. Neuromuscular disturbances, when present, are bilaterally symmetrical.

The various disturbances listed above are not all seen in every affected animal; rather, there are syndromes, in which a selection of these clinical features occur. These syndromes differ in their finer details, but can all be related to five main physiological disturbances:

(1) of metabolism, leading as a rule to inanition and loss of muscle mass, but occasionally to obesity, and causing abnormal intake of water and sodium chloride, without loss of appetite;

(2) of motor function, with dysmetric ataxia of the limbs, but without disturbances of body righting reflexes;

(3) of sensation, causing compulsive rubbing or nibbling of certain parts of the body, without evidence of skin disease, and elicitable by pressure on deep tissues;

(4) of behaviour, with signs of emotional instability, anxiety, and confusion;

(5) of autonomic nervous control, with tendency to tachycardia and cardiac arrhythmia, distortion of normal alimentary mobility, and inability to maintain bodily homoeostasis under conditions of moderate exposure.

Some but not all of these disturbances can be related to specific neurohistological findings, which are described in Chapter 7.

Atypical symptoms have occurred in scrapie-affected kinships often enough to be accepted, for all practical purposes, as part of the scrapie complex. In several families an individual atypical syndrome has appeared in consecutive generations. A few such syndromes are described below.

Posterior Limb Palsy or Paraplegia

Hind-limb palsy is a very characteristic acute symptom, in which the hind limbs suddenly become weak and the animal is unable to walk. In females this often occurs just before or just after parturition; it progresses so rapidly that within 10 to 14 days the animal cannot rise on its hind limbs; its tendon reflexes are still present, but weak. Movement of the fore part of the body is virtually unaffected; hence the sheep sits up on its haunches like a dog. In males it may develop at any time of the year, but is most commonly seen in the late summer and autumn, and is usually attributed to fighting among a group of rams. Very occasionally a somewhat similar palsy confined to the forelimbs of young rams has been noted. The palsy is frequently of sudden onset, and thereafter is always slowly and relentlessly progressive. It does not show any abatement after 48–72 hours, as commonly occurs with traumatic injuries associated with acute spinal cord oedema of the Schiff–Sherrington type; response to pin-prick and plantar pressure are unimpaired, and no spinal deformation can be detected by careful palpation. X-ray examinations have not been practicable.

These palsies almost always occur at the earliest part of the age range of clinical manifestation (see p. 69) and progress rapidly, without wasting of the musculature. Some nibbling response is usually present. One ram, a Royal Show champion under test-mating (see Chapter 6), developed this syndrome at age 1¾ years and two of his daughters out of scrapie-affected dams were similarly afflicted.

Acute Myasthenia

Some animals, from scrapie-affected kinships, show a rapidly pro-gressive general weakness and disinclination to walk far, become unable to rise from a normal resting position of dorsal decubitus, and when lifted to their feet collapse to the lying position within a few minutes. They are almost always in good general physical condition, and are frequently fat. Appetite and reflexes are unaffected; the animals remain alert and, with careful nursing, they survive some weeks. They do not respond to neostigmine injections.

Defective Vision

Some 5% of our cases of undoubted scrapie, this diagnosis being based on clinical and pathological criteria, have shown defects of vision. Indeed, this has often been the presenting sign, and when it is present the animals are always kept under observation as probable cases. The pupillary light reactions are normal, and the pupils of normal size, but the menace reflex is absent. Eyeball movements are normal, and nystagmus has never been seen. The eye and its fundus on ophthalmo-scopic examination are normal; there is no papillary oedema or anomaly of blood vessels. At autopsy the eyeball and optic nerves are normal, there is no constriction of the optic canal or any gross macro-scopic change in the optic nerves, the geniculate bodies or calcarine cortex of the cerebrum. On microscopic examination the retina is normal, and no changes have been detected in the central optic pathway; however, the lack of adequate precise knowledge of the structure of the geniculate bodies and the visual cortex of the sheep has prevented thorough study.

Neurological blindness of this kind is very uncommon in our lowland sheep flocks without access to bracken; the 'bright blindness' syndrome (319a) of certain bracken-infested hill areas may be excluded. Indeed we have never encountered a case of this type of visual defect apart from scrapie in our 25 years' study; the transient blindness of pregnancy toxaemia is quite different. Barnett and Palmer (17) have reported blindness in two sheep with histopathological evidence of scrapie, but they observed retinal damage, which is not a feature of primary scrapie blindness.

Epilepsy

Very rarely scrapie may present as motor seizures of grand-mal type, with characteristic evidence of scrapie developing in 2–3 weeks. Epilepsy is virtually unknown in sheep; no syndrome such as the hereditary epilepsy of Brown Swiss Cattle (12) is known to us.

Slow Progressive Inanition or Ill-thrift

This syndrome, for which no nutritional, helminthological, parasitological or other pathological cause can be detected, occurs in rams and older ewes. In rams, affected in the usual age range of scrapie, careful and persistent examination may reveal a nibbling reflex, but evidence of ataxia is commonly equivocal or slight. A firm diagnosis depends on excluding other causes at autopsy, and on neuropathological examination, which must include the mid-brain; but a longer period of clinical observation may be decisive. In ewes older than 5–6 years, this form of the disorder usually manifests in early and late autumn, as pregnancy progresses. Other signs of scrapie are so inconspicuous as to be disregarded, and the diagnosis depends upon full neurological examination, which reveals the loss or vacuolation of neurones in specific locations.

Segmental Rubbing

The rubbing so characteristic of most cases of scrapie may in rare instances be strictly localised, e.g. to an annular band 5–10 cm wide running vertically across the thorax and/or lumbar region, or to a narrower band, 2–4 cm wide, on the head, usually just dorsal to the eyes.

IV. The Age of Onset, Course and Outcome

Onset of the disorder is usually between the ages of 2 and 5 years, with 5–10% of all cases manifesting after 5 years of age; that is, the disease is clearly age-related. We have not seen any cases in animals less than 1 year old, such as Joubert *et al.* (165) have reported in southern France. One possible instance in the 1930's in England was recounted to me, but no recent instance has come to my attention; such early cases in adolescence, diagnosed as scrapie, may represent a different kind of disorder, which in the absence of pathological confirmation must remain indeterminate. It is not infrequent for some progeny to show the disease before their parents are affected, but not before the age of 2 years. Both sexes appear to be equally prone, with rams manifesting at a slightly younger age.

The course of the disease is almost always slowly and relentlessly progressive, with death in 3 to 6 months. Out of 3,000 cases, we have had three instances of remission at age $2\frac{1}{2}$–3 years, but with subsequent relapse proceeding to death at age $4\frac{1}{4}$–$4\frac{1}{2}$ years. In each instance the initial clinical syndrome gave rise to some uncertainty, although the final phase was typical.

The duration of the illness varies considerably between individuals, for reasons not yet understood. Earlier claims that a first pregnancy as a lamb (226) or the starving of sheep, will hasten the development of the disease have not been confirmed in our trials. The disease has appeared more frequently in two periods of the winter months in Britain, which coincide with late pregnancy from mid-October to December, and with parturition and nursing the lambs in January to March; but I suspect that this may, in part, be due to the removal of the less desirable or productive sheep from the flock in late spring and early summer, when potential cases of the disorder in an early preclinical stage will tend to be culled.

The outcome of the disease is a slow decline into a physical state in which the animals may still feed themselves, but contrary to the usual behaviour of prostrate sick sheep, they may live for many days with only a minimum of food and water. This phase is usually associated with extensor rigidity of the limbs of the side on which they lie, a position from which it is very difficult to move them.

Death is usually at night, especially if the animal is exposed to cold rain, or following some emotional disturbance such as pen changing, if it is housed. Pneumonia is uncommon, and aspiration of ruminal contents is very rare. There is no evidence to suggest terminal enterotoxaemia, against which all our flocks have been vaccinated as a routine procedure. If the disease develops in the late pregnancy of a twin-bearing ewe, the terminal cause of death is frequently toxaemia of pregnancy associated with the disturbed and inadequate food intake consequent upon the behavioural disturbances. As already noted, death by asphyxia after being 'cast on the back' is a special risk in obese animals in full wool.

No effective means of preventing the disease, delaying its course, or altering its outcome are known. We have tried, in addition to good nursing and feeding, each of the following treatments on one or two animals, but without any sign of modifying the expected course of the disease.

(i) Vitamin B_1, given subcutaneously as 1 g daily of aneurine hydrochloride in aqueous solution for 2 months.

(ii) Vitamins B_1, B_2, B_6 and nicotinamide given as a combined preparation (Becosym) intramuscularly, providing 20 mg, 8 mg, 8 mg and 80 mg, respectively of the four vitamins twice daily for 6 days.

(iii) A supplement of high-quality protein, minerals, and accessory food factors fed in the form of dried liver (Hepavite) and spray-dried whole milk powder at levels of 37 to 75 g and 200 to 300 g respectively for 40 to 50 days.

(iv) Vitamin E as α-tocopheryl acetate given subcutaneously as 350 mg in arachis oil daily for 32 days, and thereafter 200 mg daily by mouth for 20 days.

(v) Vitamin B_{12} given subcutaneously in doses of 2 mg daily for 4 days, and thereafter 250 μg daily for 15 days.

(vi) Desoxycorticosterone acetate given subcutaneously as 15 mg in oily solution daily for 35 days.

(vii) Progesterone given intramuscularly as 20 mg in oily solution daily for 50 days.

(viii) Adrenocorticotrophic hormone given subcutaneously in a dose of 40 mg I.U. twice weekly for 3 weeks.

(ix) Three 5-day courses of British anti-lewisite (BAL) with an interval of 2 weeks between courses.

(x) Neostigmine in single doses of 2·5 to 5·0 mg intramuscularly. This may sometimes provide a slight, very transient, improvement in the early stages.

(xi) We have also conducted a controlled trial in two groups of ten affected ewes in an on-farm trial, using 300 mg of α-tocopheryl acetate daily against a placebo of the vehicle only, without effect over a period of 2 months.

V. Diagnosis of Natural Scrapie

There are at present no simple specific tests for the presence, past or present, of the transmissible agent (or slow 'virus'). The chromosome number (46) and pattern is normal. The erythrocytes and white blood cells are normal in numbers and morphology. The serum proteins appear to be within normal limits. The ratio of high K^+ and high Na^+ erythrocytes is variable. No pattern of haemoglobin specificity for the A, B and X forms has been detected, while the evidence on blood group substances in sheep remains confused. No specific disturbances of the chemical constituents of the blood, urine, saliva or semen have been detected, and the transmissible agent is not present in these materials. In practice, the agent can only be demonstrated by long-term animal inoculation. The cerebrospinal fluid may contain a small increase of cells, but such increases, and changes of protein levels, are inconstant. The report by Field and Shenton (117) that the macrophages of sheep with scrapie showed an enhanced mobility in an electric field in their macrophage-electromobility test was therefore welcomed as potentially an important advance. Alas, careful evaluation of the test by Fraser and his colleagues (121) has failed to confirm that it is of any value in detecting or confirming scrapie, and indeed its suggested value for screening for potential ill-health in human medicine is doubtful.

Electroencephalography (EEG) has been used extensively in analogous human disorders, including kuru (79, 80). In keeping with the predominantly cerebellar pathology in kuru, the EEG changes are mild. Without any background information on the normal EEG of the sheep, we have considered this approach to offer too uncertain an outcome, and have not used it. Joubert *et al.* (166) have published such a recording in their 'Juvenile encephalopathy with vacuoles' in three animals; they claim distinctive changes, but offer no evidence that the normal EEG in sheep of the same age, breed and background has been established. The more recent report by Court *et al.* (84) is open to the same criticism.

We are thus forced to continue to define the disease in terms of its clinical signs and pathological features, with an artificially transmissible agent when its assay is practicable.

Chapter 5

Epidemiological Studies

[*Editor's note*: the data presented in this chapter, and in Appendix A, come from Parry's 25-year collaboration with the members of the Sheep Development Association (SDA). This body, which survived the shock of Parry's sudden death, consists at the time of writing of some 12 major breeders of pedigree Suffolk flocks, widely distributed through England and Wales, comprising about 1600 breeding ewes.]

I. The Natural Pattern of Occurrence of Scrapie in Sheep

When I was drawn into the tight world of professional sheepmen in East Anglia in the early 1950's, it soon became apparent that the accounts of scrapie in the veterinary literature gave a very simplistic, superficial and inexact account of the disease as it had been encountered in the previous 30–40 years in Suffolk and Norfolk. Some considered it to be infectious, but many did not. The primary need was to establish exactly what scrapie disease might be, what clinical forms it might take, how and where it occurred, how it might be disseminated, whether by an infection or through breeding, and how controlled. Such a task is only possible with the goodwill and free collaboration of the shepherds and flockowners controlling a large semi-closed interbreeding sheep population over at least a decade, or better, over a quarter of a century. The local breeders of registered Suffolk sheep undertook in March, 1952, to collaborate in such an investigation, based on linked recordings of health and parentage specially designed to ensure a high degree of accuracy and reliability.

The published studies of the disease have been generally inadequate to reveal the basic epidemiological facts which are needed for any proper understanding of the mode of dissemination of scrapie. Indeed it is only in the last few years that collaboration with this group of sheep breeders, with its standardised record-linkage scheme, has gone some way to provide this crucial information, to highlight the importance of kinship and inheritance, and to show how a recessive gene may produce outbursts of the disease simulating infectious epidemics (106) and how sporadic and capricious outbreaks may be caused, without postulating the action of a communicable virus-like agent.

These epidemiological field studies began in 1953, and expanded to pedigree breeder groups of 11 breeds, with flocks sited in many parts of Great Britain outside East Anglia, and linked with a breeding and hospital flock of several hundred animals maintained as a farm enterprise at Oxford. With access to this source of recorded affected and normal sheep, studies of the clinical disorder and its pathology have been conducted in collaboration with colleagues in Oxford and at the Maudsley Hospital in London. The following pages set out the main conclusions of the Oxford studies on scrapie over 25 years.

In 1950 detailed and reliable information on scrapie was very difficult to come by in Britain. Recent published accounts were largely based on laboratory studies of the experimental disease. The practical experiences of the period 1750–1850 had been lost from the written literature of the subject and from the 'folk-memory' of sheepmen, at least in the south of England. The disease was known to be incurable, and unpredictable in its occurrence and course; and its presence carried a peculiar stigma of social disapproval. Most knowledgeable flock-masters and shepherds had deliberately erected a barrier of silence, and refused to admit to any knowledge of the disease; only the inexperienced talked openly or sought advice. Owners, many highly respected, often clearly indicated to their shepherds that they did not wish to be informed about any possible cases in their own flocks, which were to be put down and unobtrusively buried. They were then free to say they knew nothing about the disease. Only a chance invitation to take part in a public 'Brains Trust' on Sheep at Bury St. Edmunds, West Suffolk, in 1951 led to an avalanche of private disclosures which provided an opportunity to gain the confidence of East Anglian flockmasters and shepherds so that they could talk freely and answer questions on the disease, which was then of major concern to many flockowners. It soon became clear that scrapie took several clinical forms, which did not appear to bear very much relation to the published neuropathological lesions; that the cause and control of the disease was hotly debated, since the occurrence and attack-rates in individual flocks showed wide variation, the course of the disease in a flock was unpredictable, and the consequences of introducing an animal which later became affected were apparently quite capricious. From this experience some inferred that the disease was infectious, while others still considered that there was a strong hereditary influence. Yet no single breeder knew enough to construct a plausible practical means of control. Thus the paramount need appeared to be to establish what scrapie was, where it occurred, and equally where it did not occur, how cases in one flock might be related by contact or by pedigree with those occurring in another flock, and if possible to relate clinical observations with pathological lesions in physiologically eloquent locations.

To achieve reliable data of this kind, the special interest and full collaboration of influential flockowners and senior shepherds of a major breeder group, with an interbreeding population, was obligatory, since no single flock in Britain could hope to provide the population size required. A breeder group of the Suffolk breed of 32 flocks, with 5000 ewes, entered into a collaborative research and breed-development study in 1952. At that time very few of these flockowners were keeping records adequate for a precise epidemiological investigation, and in most other breeds full pedigrees were not even kept, although commonly claimed. The first stage was to evolve a system of detailed management and recording within the group, so that reliable parentage and incidence data might be collected. With the help of several experienced shepherds a system was evolved which fitted into the shepherds' daily and seasonal work-pattern, was acceptable and of interest to them, and was therefore least likely to attract human errors and omissions. This system has been in use for the last 20 years, and has been applied in 12 breeds. Its success may be gauged by the presence of our field book in the shepherds' breast pockets. It is the data collected by this voluntary system which form the basis of the following account.

II. The Collaborative Research on Scrapie in Suffolk Flocks

The Background

Scrapie has been particularly subject to the special and well known equivocation exhibited by pedigree animal breeders when dealing with hereditary complaints. They do not wish to publicise any defects in their own animals, and often go to considerable lengths to mislead the casual enquirer, and to prevent the real facts from emerging, as May (199) remarked in Germany a century ago. It is an ancient craft, this weaving of a web of deception to protect one's cherished flock from outsiders' knowledge. The earliest account of 'goggles' from Wiltshire to the Bath Society in 1788 was from an anonymous 'Wiltshire gent' (10). My present informants still prefer to remain anonymous, or to pass on crucial information in indirect ways.

The relationships and bonds between stud pedigree breeders are much the same throughout the Western world, and have probably not changed significantly over the past two centuries, since Bakewell set up the rules for the hiring of Dishley Leicesters (282). Similar economic and social ties and constraints still apply in this free market economy of many small independent producers, which remains free of centralised bureaucratic direction in the English-speaking and most western European communities.

The question of a legal warranty as to the subsequent performance of such animals offered for sale may also arise. Around 1800, 'goggles' was causing such havoc in Wessex that the sellers of sheep subsequently affected were, by general public pressure, obliged to stand the loss (95), although there was no formal legal requirement. The tradition has now disappeared, although some sellers may make refunds of purchase money, or mitigate losses in other ways. In France (129) contractual obligations were more clearly defined. One must act with great discretion if one wishes to ascertain the real facts regarding scrapie, and respect at all times the point of view and interests of flockmaster and shepherd. For it is not merely the legal and monetary liability; a greater blow is the loss of 'face' among other breeders if it becomes public knowledge that one's flock, the life interest of a good shepherd and a good flockowner, is tainted with scrapie. Indeed such was the attitude in East Anglia in 1950 that I, as a known qualified veterinarian, was suspect on the grounds that other farm staff and neighbours would assume there must be 'disease trouble' in any flock I was seen to be visiting. It was essential that by repeated visits and by long enquiries and discussion on non-veterinary matters of sheep breeding, it became known that I was not really a veterinarian but merely an odd eccentric, interested in all aspects of sheep breeding, who had some practical aid to offer. After a year or two I became accepted in this role, and through the shepherds' national 'grape-vine' of news exchanged at shows and sales, I became acceptable over much of Britain, although some important breeders would not collaborate freely with me, nor, I have reason to believe, with any one else. On my side it was essential to identify myself fully with the aspirations and rewards of the shepherds and flockowners in their show and sale successes, and to help these by any means which might come to hand.

Any investigation purporting to be aimed at a credible ascertainment of the scrapie position in any flock or population must show that it has taken heed of, and succeeded in surmounting, these problems of personal relations, social anthropology and agricultural economics. Absolute confidentiality of all information is a *sine qua non*, and no fear of untoward consequences arising from disclosure must be allowed to occur. It should be noted that the period 1950–1970 in Britain was a peculiarly fortunate one in which to attempt a study of this nature. The organisation of the British pedigree (i.e. pure-bred) sheep industry had enjoyed a period of stability and slow evolution for two centuries (15, 16). The best shepherds were steeped in a tradition of flock management which was often reinforced by intermarriage between shepherding families. This tradition had been shaped by a variety of misfortunes, resulting in empirical adjustments of flock management according to local circumstances without much help from biological or veterinary science, apart from the use of new prophylactic inoculations. The owners respected the views of their shepherds, and rarely overturned their shepherds' decisions. The owners were still deeply interested in their sheep, as in the 18th and 19th centuries; firstly for economic reasons, since the use of sheep on the break-crops of the arable rotation, e.g. the Norfolk 4-course rotation associated with Townshend, was still necessary, and not yet disturbed by the introduction of vegetable crops for canning etc., and chemical sprays for weed control. Secondly, the role of exhibiting successfully at the local County Agricultural Shows, handed down from the annual public sheep-shearings and discussions of the early 19th century at Holkham in Norfolk (1775–1820) and Woburn in Bedfordshire, and carried on by the shows of the Royal Agricultural Society of England from 1839 and the Highland and Agricultural Society of Scotland from 1785, and later County Societies (216), still exerted a substantial influence on owner-management attitudes. Success at local shows was a matter of intense local pride, carrying a strong social status, apart from its marketing and advertising advantages.

By 1970 these attitudes were changing. The demands of agricultural shows during the 1950's had kept many shepherds away from their flocks for much of the summer, e.g. in East Anglia for 3–4 days each week from mid-May (the Hertfordshire Show) to early July (the Royal Show) and even until tupping commenced on 6th August; but the social and economic importance of showing had declined, and many flocks had given up showing because of increased costs and man-power demands. The social importance of show successes in maintaining farm and estate morale had passed, and with it the close interest of many English owners in their flocks. The diligence of shepherds had also suffered erosion from the late 1950's. Television and motor cars became generally available, and offered alternatives to the single-minded interest of the older men in the sheep and their welfare. The work, which had formerly occupied 7 days a week, was now cut down to the 5 or $5\frac{1}{2}$ days demanded by their women-folk and encouraged by public attitudes.

As the 1970's proceeded the impact of legislation and urban development on the management of farms deflected the attention and energy of owners away from the breeding of their livestock, while social changes in staff attitudes militated against the careful and continuous observation of their sheep by shepherds, who found their once-privileged status in the hierarchy of the countryside seriously diminished. These changes have led to a marked reduction in the number of flocks with a farm organisation capable of undertaking the more complex breed improvement programmes.

Design of the System of Flock Recording and Management

The historical evidence outlined in Chapter 3 indicated that scrapie might be controlled by hereditary as well as infectious components, that it was a disease of middle age, with variable age of onset, and that it was capricious in its initial occurrence and subsequent course in any sheep community. In 1950 there was no method of detecting liability to, or confirming the presence of, the disease in the living animal; and with animals of unknown ancestry there is still no method. There were also no published accounts which gave any recent quantitative data on prevalence, ancestry, effects of contact between affected and un-affected individuals and communities, or on age of clinical manifestation, the best account of these matters being nearly a century old (199).

In order to arrive at valid conclusions regarding the natural history of scrapie, a method was needed of assembling accurate health, production and parentage records within closed or semi-closed breeding groups over several generations, i.e. a family or clan health record linkage scheme. The assembly of data on suspected hereditary ailments first manifesting in middle-age or later raises special difficulties, due to the dispersal of animals by sale and the loss of individuals not found suitable for breeding and the customary absence of written records made at the time.

In 1950 no sheep breed society or similar organisation was assembling such data, and very few individual flocks were keeping records sufficiently comprehensive to reveal the possible role of heredity in scrapie, or in any other disorders not obvious at birth or in early infancy. The lack of (unsought) evidence for heritability was taken as evidence against it. It was assumed that scrapie must be a conventional infectious disorder, since Cuillé and Chelle (87) had been able to set up a similar syndrome in laboratory animals by inoculation.

To meet the needs of this situation, a system was devised, based on daily recordings by the shepherds, in bound tamper-proof field books, of the required items of information. All sheep are identified by two permanent marks of identification, e.g. two ear tags, a tag and ear notches or tattoos, or two horn brands. At the beginning of the flock year an inventory of all sheep in the breeding flock is made in the field book, with the allocation of ewes to each ram and to the enclosure in which the matings take place. Wherever practicable, as in all the flocks supplying important records, the ewe's identity (year letter and number) is stamped on the fleece over her upper mid-thorax, using marking fluid 'paint' brands 8–10 cm high, and her ride* number stamped on her upper neck. This number is also placed on the ram, which carries raddle (red ochre) on his sternum, either applied directly or in a sire-sine harness†. These fleece brands allow identification from a distance, and provide a check that all members of mating groups remain in their respective enclosures, which are properly fenced, with a sheep-free zone between groups. The system

---

* A 'ride' is a mating group—i.e. a number of ewes served by a single ram.

† A device strapped onto a ram, producing a distinctive mark on the backs of the ewes which he has served.

also allows recording of mating dates which, with the recorded lambing dates, provide a valuable check on paternity, if any query arises,
even years later. At parturition the lambing ewes and their offspring are placed in individual pens for 2–4 days, and the lambs
identified by tags or notches before leaving these pens; full details are
entered in the shepherd's field book, in two different sections, to check
their accuracy. Details of all cases of ill-health and deaths are recorded
on the day they occur.

The data in these field books are transferred to large bound flock
registers to provide an easily ascertained record of the parentage and
the life history of each individual, and as a check on errors and omissions.
Twice a year, in autumn and spring, each flock has been visited by the
author in order to check the records and the health of the flock, to
discuss any matters of common interest, and to decide on breeding
policy.

Diagnostic Criteria

The clinical features are so distinctive (see Chapter 4) that well-
informed sheepmen rarely fail to notice and report a case correctly,
although those without experience frequently fail to do so, especially
in long-woolled sheep. Of about 3000 cases, 2100 have been observed
by the author and 1900 autopsied by him, of which 1040 were subjected
to a full neuropathological examination carried out 'blind' (Appendix A).
Any animals with doubtful or atypical clinical signs have been brought
to a hospital farm at Oxford and kept under observation, often in
paddocks under direct view from the residence, until a definite diagnosis
has been established.

The diagnosis of 'affected' or 'unaffected' has been based on four
considerations:

(1) The clinical signs, recorded by an observer known by us to be
competent and in a position to make continuous observations of the
progress of the illness.
(2) The presence or absence at autopsy after euthanasia, i.e. before
post-mortem artefacts develop, of any gross macroscopic abnormality
other than small changes in the brain, endocrine glands and skeletal
musculature. The main organs are weighed and compared with our
normal control series for different breeds, ages, sexes and times of year.

(3) Detailed examination of the brain and upper cervical cord after preliminary fixation within 30 minutes of death, first macroscopically on cutting coronally at 12 levels, and then microscopically at nine levels after using five or six staining methods, the evaluation being carried out 'blind' by one of three observers (for details of the lesions sought see Chapter 7).

(4) The presence or absence of syndromes with which scrapie might be confused. None of our flocks is located in an area where louping-ill encephalitis is endemic. Cutaneous scabies ('scab') was absent from Britain between 1952 and 1975. Listerella and Aujesky's encephalitides have not been encountered. Rabies and Borna disease are absent from the United Kingdom. Certain intracranial space-occupying lesions occur sporadically, such as coenurosis of the brain (gid) and the acute cerebral oedema known as cerebrocortical necrosis, which are readily distinguished clinically. One animal with a diverticulum of the fourth ventricle involving one foramen of Luschka, a very rare anomaly, simulated scrapie clinically but was recognised histologically as non-scrapie. Intracranial tumours and cerebral vascular anomalies in sheep are extremely rare. No brain tumour or vascular anomaly has ever been observed in our series. A small benign adenoma of the pituitary was found once. Behavioural disorders due to arteriosclerosis or trauma are unknown in these sheep.

Assessment of the Accuracy of Diagnosis

The degree of accuracy achieved may be judged from the results of a series of 437 consecutive autopsies, on 336 affected and 101 unaffected animals, the latter serving as contemporaneous controls obtained from the same birth cohorts and flocks. Of the 336 clinically positive animals 334 were adjudged fully positive on histopathological grounds, with the other two probably positive. Out of 101 unaffected animals, 93 were entirely negative on their histopathology, with six probably and two definitely being positive. As all eight animals were older than 5 years, these probably represent cases of late manifestation, in which the extent of the neuropathological lesions was not yet extensive enough to produce deficits of function sufficient to cause clinical signs. In only three animals under suspicion of scrapie, but with atypical signs and showing a remission of the progress of the disease, has consideration of the pathological evidence not allowed a definite diagnosis. At present no other syndrome simulating scrapie has been recognised, and these animals were therefore considered to be aberrant forms of the disease for purposes of determining breeding policy.

Flock Management, Breeding Practices and Sire Selection

In 1950 a substantial number of lowland sheep were still close-folded in enclosures, 50 to 150 metres square, of hurdles (now replaced by electri-fied wire mesh netting) on arable land, which allowed individual mating groups. They lambed in the open, for which, if lambing was between December and March, substantial lambing yards were constructed each year on a different site. These arrangements have gradually been abandoned in favour of extensive grazing on leys, with less reliance on root and arable forage crops, with lambing in permanent farm buildings, the use of which was previously eschewed on account of the possible build-up of infection and the need to take the manurial residues of the flock round the farm with the rotation. With availability of electric light accurate recording is now practicable. On most lowland farms the sheep have been kept, mated and lambed in fields of permanent herbage and grass.

In the hill breeds the ewes spent most of the year on the high ground, being brought down to the low ground in bye-enclosures just before lambing, and returned uphill a few weeks thereafter, in many cases to hill grazings held in common with other flocks. The lack of sufficient secure enclosures prevented individual mating groups, except on exceptional properties. A few élite breeders were able to achieve this, but full parental pedigrees remained difficult to organise, and most breeders were only able to offer data on a lamb's sire based on the raddle colouring of its dam, since fleece-branding of long-woolled breeds is unsatisfactory, and neck-bands are impracticable for most flocks. The hill breeds are bred pure for flock replenishment for the first two or three gestations; the ewes are then mated with rams of other breeds to produce half-bred stock, which are sent to the lowlands as commercial sheep, until they themselves are sold off the hill after four or five lamb-crops to lowland sheep-keepers, when they are lost sight of.

The selection of rams was based firstly on phenotype inspection for the breed characters as established at Agricultural Shows, taking into account unacceptable faults or weaknesses; and then the final selection would be made wherever possible on the production records of the parents (notably the breeding performance of the sire) and their fore-bears. Breeders tended to develop their own preferred type, as far as financial constraints allowed. Some rams were usually retained from within a flock. The more highly regarded the flock, the more this tended to occur; but most breeders selected a third to a half of their stud rams from other flocks, approximately one-half being related to their own stud rams and one-half unrelated, as far as possible, but related to recent oustanding sires of other breeders. The females were selected largely on phenotype inspection, and on the lifespan of their dams. This system prevented close in-breeding, and in general pro-tected breeders from the troubles of sub-fertility and reduced viability (so called 'lack of constitution') and inability to thrive and survive. It may be termed a loose system of line-breeding. It proved satisfactory as long as the breeding group was large enough, say at least 2000 females, although more are desirable. In several breeds of smaller size, the unfortunate use of show prize-winners without sufficient attention to life-time performance of their forebears and progeny has led since 1950 to the virtual disappearance of these breeds.

Any investigation of heritable characteristics requires, to be successful, access to the group of flocks within a breed from which the élite breeders will normally select their rams, with the whole group voluntarily keeping records to a standard pattern. The rams are mated with 10–50 ewes each year, the ewes rarely being mated with the same ram for more than two of their five to six gestations. Each flock is self-contained for females except for an occasional purchase. Thus each breeder group is virtually a closed breeding population with a relatively high degree of consanguinity and extreme polygamy.

Size of Population Units and Their Age Structures, with Duration and Variable Completeness of Ascertainment

The breeding units, the individual flocks of pure or hybrid stock, varied from five to about 300 females, rarely more in the lowlands. Those observed in our recording scheme usually numbered between 50 and 300. They were organised in isolated flocks breeding their own female flock replacements, and receiving only one to three rams from other flocks each year. These flocks were registered with a breed association or society comprising 50 to 1200 member flocks, with a total female breeding population ranging from a few hundred to an estimated 10,000 or more, and which were mated once a year (the Dorset Horns excepted). In practice the effective size of the intrabreeding population of any breed was much more restricted, rarely exceeding 10% of the total registered in the breed.

The duration and completeness of observation also varied considerably, being determined by the age profiles customary for the breed. Thus in lowland breeds at least three main categories of flock age-structures occur:

(1) Flocks keeping their flock ewes until removed by death, or by 'culling' for low productivity or chronic ill-health or senility. This occurs in most Down breeds, Romneys, Lincolns, Leicesters, etc. About 40% of the flock will be $4\frac{1}{2}$ years old or more.

(2) Flocks of breeds traditionally selling all females as 'draft' breeding ewes (i.e. for possible further breeding in non-registered flocks) at $4\frac{1}{2}$ years old after three gestations, e.g. Dorset Downs, or after four or five gestations, e.g. Dorset Horns and Poll Dorsets. No sheep older than $4\frac{1}{2}$ years will be retained.

(3) Flocks traditionally selling all females as young breeding ewes at $1\frac{1}{2}$ years of age after one or two gestations, e.g. Clun Forest and Kerry Hill. In hill and semi-hill breeds of the North and Wales, all females are sold as 'draft' ewes after four gestations (occasionally five or three), at $5\frac{1}{2}$ years old, or occasionally at $6\frac{1}{2}$ or $4\frac{1}{2}$ years, depending on the 'hardness' of the hill grazings. Very few of these animals are more than $5\frac{1}{2}$ years old; most are much less.

Thus it is effectively only in the lowland breeds of category 1, or in individual flocks of other categories which have departed from their customary breed pattern, and adopted a category 1 breeding policy, that any meaningful ascertainment of the age of clinical manifestation of scrapie can be obtained. Even in this general category, the age structure of the flocks with which I was concerned varied with the breed. Normally 20–30% of the females of a birth cohort, with 1–2% of the rams, were retained to maintain the flock; the remainder were sold by $1\frac{1}{2}$ years of age, and generally lost to observation, unless special means of follow-up were used. The age structure of the flock ewes normally allowed for about 40% to be retained after the third gestation (at $4\frac{1}{2}$ years of age), and thereafter as long as no disability impaired their productivity, a few of the oldest being 9–11 years old.

The rams' life-span rarely exceeded 7–8 years, and many whose progeny were of unacceptable quality disappeared before $4\frac{1}{2}$ years of age, some by $2\frac{1}{2}$ years old. Thus the proportion of the lambs in each birth cohort which could be observed to 4 or more years of age rarely exceeded 10–15%. In a few breeds, e.g. Dorset Horn and Dorset Down, Clun Forest and Kerry Hill, virtually no breeding females were available for observation after the age of $4\frac{1}{2}$ years, except in unrecorded commercial flocks of unknown provenance.

A special attempt was made in 1964–1966 to follow the hybrid Masham females sold out of certain Swaledale and Dalesbred flocks of the Middle Pennines sired by known Teeswater rams. Some 12,000 Masham lambs were identified by ear tags on the farms of origin in 1965, and 7000 in 1966, of which it proved possible to trace and keep about 1000 under observation in commercial flocks in Lincolnshire. Scrapie occurred in these Mashams in 1968–1971, and in one instance it was possible to trace the ancestry back to a Swaledale flock known to have scrapie and to a Teeswater ram, which we had had for examination, and had already confirmed as suffering from scrapie. Normally such record-linkage is not possible under present British agricultural conditions, and was only achieved by a very unusual collaboration between the breeders, the dealers purchasing the young Masham females at auction, and the ultimate farm purchasers.

Thus the data available on the epidemiology of scrapie in Britain are limited to a very small sector of the total British sheep population of some 12 million breeding females, and will not disclose disease developing after early middle age (4–5 years), except in special circumstances. The completeness of ascertainment will also be seriously distorted if, in the selection of the young animals retained for breeding, there is any selection pressure or other bias against scrapie-prone juveniles. In practice none has been detected, and in some breeds there is a strong selection pressure favouring retention of the scrapie-prone animal (284).

Special Factors Leading to Incomplete Ascertainment of Total Scrapie Attack Rates

1. Delayed age of clinical manifestation (ACM)

During the decade 1950–1960 the first 1000 cases of scrapie observed, mostly in Suffolks, exhibited a mean ACM below 4·0 years (see Figs 5.1 and 5.2) with between 85 and 90% of all cases occurring by the

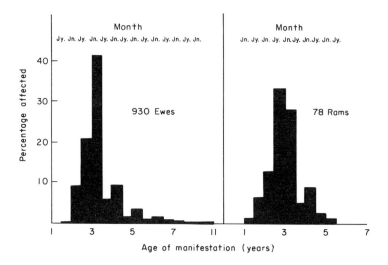

Fig. 5.1. *The age, in six-month periods, at which unequivocal clinical signs of scrapie were first manifest in 1008 sheep. Figs 5.1 and 5.2 are taken from Parry (1962) by courtesy of the Editor of* Heredity.

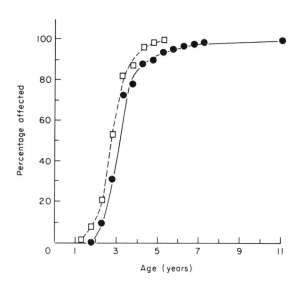

Fig. 5.2. *Curves showing the cumulative manifestation for 78 rams (□) and 930 ewes (●), based on the percentage of all animals of each sex affected by a given age.*

age of 5·0 years, males being affected on average a few months earlier than females. The second 1000 cases have shown a similar pattern. The data on Suffolks in Scotland reported by Dickinson (98), and most published reports, are in broad agreement with this.

However, should the ACM in any population be delayed to an older age, leading to a higher proportion than 10% of the total potential cases manifesting after the age of 5 years, a substantial under-ascertainment of the total scrapie attack rate could occur. Four lines of evidence have emerged suggesting that such late manifestation may indeed happen under certain circumstances, although its causation and quantitative importance are not clear. (1) Some individual Suffolk sheep manifesting at age 7–9 years have tended to have more progeny manifesting after 5 years old than had been expected. (2) In one flock, in which the attack-rate of the disease has declined dramatically, following the use of test-mated rams (see Chapter 6), the ACM now being observed tends to be 4 years and older, instead of the 2–4½ years observed between 1950 and 1970. (3) Some clinically normal elderly sheep in this flock have been found to have definite neuropathological evidence of scrapie (228), as if the rate of progress of the neural degeneration had been slowed. (4) In one hill breed flock which was being managed as a specialised inbreeding experiment, with retention of many ewes up to 10–12 years old, natural scrapie was first observed in ewes at 9–10 years old, but has been appearing at earlier ages in successive years for a decade, until now the ACM is 2–3 years. The change of ACM over a decade in this flock is so uncharacteristic that one suspects some special aetiological factor such as massive inbreeding or some environmental change. That such abrupt changes of ACM are in general an uncommon phenomenon may be fairly inferred from the consensus regarding the usual ACM expressed by so many observers over the past two centuries.

The practical importance of late manifestation is most critical in those breeds normally sold away after three or four lamb crops from the breeders' flocks; in these, scrapie with a late ACM may develop and go undetected and unrecognised for many years, when for reasons which we do not understand the ACM becomes earlier, i.e. 'anticipation' occurs in the flock as a whole.

2. *Loss by competing risks before expected ACM*

We have no evidence that potential cases of scrapie are more liable to early embryonic death, later intrauterine and perinatal mortality, or disabilities in infancy and adolescence (226, 227). However, we have observed a higher mortality than was expected, on the basis of the general flock record of the birth cohort, in certain individuals between $1\frac{1}{2}$ and $2\frac{1}{2}$ years old, which we had reason to expect would develop scrapie. These individuals are almost always well-developed physically, with good skeletal musculature, sometimes with a hint of clumsiness in turning at the trot. They are found 'cast' in small depressions in the ground which would not be a hazard to ordinary sheep, or they are found drowned in a dyke or deep ditch. In two such cases, examined histopathologically before post-mortem changes had developed, widespread early degenerative changes have been observed in the *boutons terminaux* of the granular layer of the flocculonodular lobe of the cerebellum such as occur in animals in the early stages of scrapie (23, 37). These lesions were sufficient to account for poor muscular coordination and movement defects. We therefore suspect that a number of potential cases of scrapie may be lost from similar 'accidental' causes before full clinical signs are manifest. In two flocks where such accidents have been as frequent as 10% of all cases of scrapie, we—owner, shepherd and myself—have all come to regard them as early cases of scrapie.

3. *Atypical manifestation and slow progression*

Animals suspected of scrapie which fall into this group can be very difficult to recognise. The acute palsies (see p. 66) unaccompanied by rubbing are often attributed to traumatic injury, but we know of no instance in which this diagnosis has been confirmed by X-ray or autopsy.

On one occasion, gestational twin rams of scrapie-affected parents presented with acute palsy within 10 days, one in Somerset and one in Suffolk. Again, some affected rams may show very little evidence of rubbing, and minimal ataxia; they present as inanition, which often progresses very slowly, much as is observed in rida in Iceland (222, 221).

In some breeds (especially the half-breds derived from the northern hill sheep, Mashams, mules and greyface), the clinical signs may be so unobtrusive and slowly progressive that even experienced sheepmen being shown round a flock with clearly affected ewes (stage 2—see p. 62) fail to recognise them.

4. Clinical normality with substantial histopathological lesions of scrapie

Here we merely note that this state of affairs has been observed in eight elderly ewes (228). How frequently it occurs we do not know, neither is any information available using the more subtle assessment based on early neuronal dystrophy (231) and neuronal counts in specific anatomical nuclei.

The general effect of these factors is to lead to under-ascertainment of clinical scrapie in any scrapie-prone population. However, one wonders whether sudden shifts, occurring over a few years, of the ACM to a much lower mean age, such as was noted in the experimental hill flock, or to a higher one, may not play a part in causing variations in the prevalence of clinical scrapie, such as occurred in English, German and French flocks in the late 18th century, in Scotland around 1900, and in East Anglia in the mid-20th century. Any rapid and widespread reduction of ACM would lead inevitably to a much larger number of cases being observed in the younger sheep, which make up the principal portion of most flocks.

Geographical, Agricultural and Nutritional Factors

The flocks observed have been in most counties of England from Devon to Durham, and in Scotland in the Tweed Valley, Fife and Caithness, at elevations up to 1000 ft (300 metres) above sea level, but mostly at 100–500 ft (30–150 metres), with annual average rainfalls of 22 in (560 mm) to 50 in (1270 mm) per annum, both within and beyond the range of cyclical deposition of sea water salts.

The soils and the underlying geological strata have been very varied, but the majority of flocks in 1950 were located on the calciferous rocks —chalks, Cotswold and Corallian limestones—which lie in a belt across England from south-west to north-east, from Lyme Regis bay in Dorset to Whitby in Yorkshire, with another concentration on the light sandy soils of East Anglia and the lower Trent Valley. A smaller number of flocks were on the Vale clays and 'red' soils of the West Midlands, and the boulder clays and silt fen soils of East Anglia and Lancashire. A few flocks were on riverine and estuarine alluvial soils. The hill flocks were mostly on Pennine limestones.

Apart from the hill sheep, the flocks were on agricultural holdings devoted to a mixed-cropping economy of cereals and livestock, with the sheep integrated as part of the arable rotation, although some holdings in the West Midlands had appreciable amounts of permanent grass shared between the sheep and a cattle enterprise, frequently a large dairy herd. In most instances the sheep spent most of the year grazing crops such as kales, swedes, turnips, sugar beet tops, tares, sainfoin and leys of Italian ryegrass, with short periods on permanent grass meadows.

The level of nutrition was uniformly good, with the grazings supplemented shortly before and after lambing with 'concentrate' feeds composed of cereals, protein supplements of legume seeds, e.g. peas, beans, pulses or locust bean, to give a protein content of 12–16%, with a balanced mineral and trace element salt additive at 1–2% by weight, usually containing vitamins A and D. The cereal crops on these farms were usually very productive and without evidence of trace element deficiencies, except for copper and manganese deficiency on some of the shallow soils over chalk, and boron deficiency on some Suffolk coastal sandy soils. There were no clear signs of mineral deficiency in any of the livestock on these farms. Nutritional ataxia of lambs has been observed on sandy coastal soils in two flocks, each in one year only, and has disappeared with additional copper administration to the ewes during pregnancy. In two other flocks, on Vale clay soils, nutritional muscular dystrophy occurred in lambs 4–8 weeks old, and was eliminated by supplementary concentrate feeds with vitamin E and selenium incorporated by the manufacturer. Bone growth and live weight gain of the young animals have been uniformly good, and much in excess of local averages.

Infective Diseases, Parasitic Infestations and Medicinal Treatments

As the flocks are virtually self-contained, and isolated from contact with other sheep, except for the small number exhibited at shows and those returning unsold from auction sales, it has been possible to maintain an exceptionally high standard of control of infective and parasitic disease by voluntary measures, as assessed by clinical inspection and sporadic autopsy data. The clostridial diseases have been controlled by the standard vaccination programmes against *Clostridium welchii* A, C and D, *Cl. tetani*, *Cl. chauvoei* and *Cl. oedematiens*. Contagious pustular dermatitis (orf) has occurred in limited cyclic outbreaks, sometimes in spite of vaccination, while contagious foot-rot has proved difficult to eliminate permanently. Rabies and Borna disease are unknown in Britain. Johne's disease, louping ill, listerellosis, Aujesky's disease, enteric salmonellosis and brucellosis have been absent clinically; occasional single isolated outbreaks of abortion thought to be due to *Toxoplasma gondii* have occurred, but no abortions due to *Vibrio*, *Salmonella*, enzootic or other infections have been recognised in any stud flocks, except for one instance of Vibrio abortion infection brought in, against advice, with purchased commercial ewes, which led to the immediate dispersal of the stud flock. A small number of cases of cerebrocortical necrosis have occurred, the clinical signs of which occasionally simulate scrapie.

Of other endoparasitic infestations, *Echinococcus* infestation has not been observed. Tapeworm infestations of lambs have been noted very occasionally, as has fascioliasis (liver fluke) in one or two holdings subject to flooding. Haemonchiosis, seen especially in ewes at the end of lactation, has normally been controlled by medication and grazing management. The common alimentary helminthiases of lambs have been rare under the management systems used. Lungworm infestations, seen in the early years in yearling females grazing permanent pastures also used by the adult flock, were controlled by management changes.

Ectoparasites have been virtually unknown. None of our flocks has been in an area infested by ticks (*Ixodes ricinus*). Scabies (*Sarcoptes ovis*) has been officially absent from Britain over most of the period until 1975; the sheep 'itch mite' (*Psorergates ovis*) has not been recorded in Britain, and 'keds' (*Melophagus ovinus*) and lice virtually disappeared with the use of chlorinated hydrocarbon dips in the early 1950's. Nasal bot fly (*Gasterophilus nasalis*) has not been seen or suspected, and the sheep head fly (*Hydrotaea irritans*) has not been present except in some Pennine hill areas. Mycotic dermatitis due to the fungus *Aspergillus dermatogonus* has been widespread, but of little importance except in one family in one flock, which were culled; the use of an alum dip in the autumn has proved to be inhibitory.

Biting lice, and flies, mosquitoes, simulidae and tabanids are not recognised at the present time as significant vectors of viral or protozoan pathogens, e.g. Babesiasis or other disorders of sheep in Britain. The former ravages of blow-fly 'strike' (cutaneous myiasis) have been virtually eliminated by the dips used since the mid-1950's.

Prophylactic and therapeutic products used as a routine in these flocks have changed over the decades. A summary is given in Table 5.1.

Table 5.1. *Summary of principal medicinal and prophylactic products used routinely, 1950–1978*

Purpose		Date
Anthelminthics		
by mouth and injection	Nicotine and $CuSO_4$	1950–54
against endoparasites	Phenothiazine (Thibenzole)	1950–63
	Thiabendozole,	
	Morantel tartrate (Barminth),	1963–70
	Levamisole	
	Thibenzole and Nilverm	1970–
Dips		
by immersion against	Arsenicals	1950
ectoparasites	Benzene hexachloride	1950–60
	'Dieldrin' products	1960–70
	Organo-phosphorus	1972–
Vaccines		
by injection to protect	Clostridial enterotoxaemias	
against microbiological	'Pulpy Kidney' *Cl. welchii* Type D	1950–55
agents	*Cl. welchii* A, C, D and tetanus (4 in 1)	1955–65
	8 in 1	1965–
	Contagious pustular dermatitis	1950–
	Contagious foot rot	1970–
	Enzootic abortion—occasionally	Rare
	Dictyocaulus (lungworm)	No
Antibiotics etc.	Sulphamezathine by mouth	⎫
	(for coccidia) and by injection	1950–
	Penicillin etc. for acute	
	infections especially post-	Sporadic
	parturient. Streptomycin	⎭
Nutritional compounds		
by injection or by mouth	Vitamins A and D supplement	1950
	Copper against enzootic ataxia	
	Selenium and α-tocopheryl acetate	Occasional
Group food additives	Antibiotics–	No
	Steroids	No
	High Copper	No

Chapter 6

The Recessive Gene Hypothesis

[*Editor's note :* The officially accepted view, on which agricultural policy in several countries is based, is that scrapie is acquired by infection. Parry believed that the natural disease was familial, and that infection in the field rarely, if ever, occurred. It is obvious that the correct policy for the control or eradication of scrapie depends upon which of these views, if either, is correct. During his lifetime, Parry experienced some difficulty in making his voice heard; and the debate, when there was a debate, was not always conducted on strictly rational lines.

If Parry is eventually to be proved wrong, it will have to be shown either that his observations were faulty or that the arguments which he based on his observations are fallacious. The reliability of Parry's data has not, as far as I am aware, been questioned. His conclusions from the data, as presented in his 1962 paper (276), were criticised in a paper by Dickinson *et al.* in 1965 (101). Parry did not accept their criticisms, and the debate lapsed. Parry persisted in the belief that natural scrapie was normally transmitted by an autosomal recessive gene, and his opponents persist in believing that it is spread by infection, the spread being partly controlled by genetic factors. The difference between these views may appear small, but it is real, and is of real practical importance. It is therefore rather disturbing to read, in a recent publication (5) on the theoretical problems in the transmission of spongiform encephalopathies the following words: "genetic control of susceptibility to scrapie has been well established (Parry, 1979)". This is precisely what Parry did not establish. Equally strange is the fact that in a recent article on the genetic aspects of scrapie (99), Parry's work is not mentioned.

The last public confrontation was when Parry published an account (228) of how the disease had been eliminated in a number of scrapie-ridden flocks by simple breeding techniques. This elicited a reply from

Kimberlin (172) supporting the contagion theory by drawing attention to three series of observations (52, 100, 155) all pointing to the reality of 'lateral' spread by infection. In his rejoinder (229) Parry expressed his reservations in accepting this evidence at its face value; and though not denying the *possibility* of spread by contagion, remained unconvinced that this was the *usual* method of spread. I am not competent to assess the merits of the arguments used on the two sides, or the probability that both sides had missed some essential clue. That the problem has no simple answer is indicated by repeated failures to demonstrate contagious spread from animals with experimental disease (235, 5).

The section of this monograph dealing with the recessive gene hypothesis was to have been the longest and most detailed of all; but it was never written. What follows here is a short statement, drawn from what was originally one of the introductory chapters. The evidence on which the theory is based is summarised in Appendix A. Dr. Gerald Draper, who as a statistician and geneticist collaborated with Parry from 1962 onwards, has contributed the following comments.

"There is no doubt that the obvious interpretation of much of Parry's data and, in particular, of the 'reciprocal matings' described in Table 1 of Appendix A is that the occurrence of scrapie is in some way determined by a recessive gene. Various criticisms have been made of Parry's analyses and his conclusions (101, 172), and there are certainly problems in reconciling his results with those of other workers. Some of these arguments are summarised below.

Parry was convinced that under natural conditions scrapie was normally *caused* by a recessive gene. There are in fact two separate controversies concerning Parry's theory.

First, there is convincing evidence (52, 100, 155) that scrapie can be transmitted by contagion. Parry, however, insisted that many of the apparent instances of contagious spread were in fact attributable to a sudden increase in gene frequency and that this could be explained by the fact that these results were obtained by workers who did not know the previous incidence of scrapie in the flocks from which these animals were derived, and were not therefore in a position to comment on his genetic theory; in particular they did not know the detailed pedigrees of their breeding rams and ewes, which Parry regarded as essential. His critics, on the other hand, argued that Parry's hypothesis implied that a large proportion of sires were heterozygotes and that this in turn implied an impossibly high gene frequency in the population. Parry countered *this* by claiming that rams which carried the scrapie gene had, in early life, highly desirable breeding qualities and that there was therefore selection pressure in favour of the gene (226, 284, 109). This claim is repeated in the present work; unfortunately, the evidence so far available is insufficient to establish the point beyond cavil.

The second controversial aspect of Parry's results concerns the question of whether the postulated gene *causes* scrapie or whether it controls susceptibility to an infectious agent. Parry believed the former and argued that the agent had never been found in non-affected animals. He also argued that though susceptibility to infection could be genetically controlled it was unknown for such control to be determined by a *single gene*. In fact one of the most astonishing features of the controversy is that Parry had produced seemingly incontrovertible evidence that there is a single gene which controls the occurrence of scrapie. We would argue that, whatever the mechanism, this is a discovery of considerable scientific and practical importance and should have been recognised as such.

There are two questions concerning Parry's analysis which should be dealt with here.

(i) It is possible that in some of his earlier work the argument concerning the genetic hypothesis was at least partly circular; for instance, some of the observations used in allotting presumed genotypes to certain animals seem to have been used as evidence of the validity of the hypothesis. The crucial test-mating experiments, however, are free of circularity in the analysis.

(ii) The estimates of manifestation level given in the papers by Parry and Draper (225–228, 106, 107) do not take into account the animals lost to follow-up through culling or other causes. It is in fact difficult to see how this could be done. It may be that the method of analysis results in an underestimate of the manifestation level, and hence leads to an overestimate of the degree of certainty that the test-mated animals do not carry the scrapie gene. However, the general conclusions drawn by Parry, his method of obtaining them, and the practical consequences, are not affected by this difficulty."]

I. Epidemiology: Facts to be Accounted For

From the mass of data collected over 25 years in the manner described in the previous chapter, a number of important generalisations have emerged.

(i) The annual clinical attack-rate in a flock may reach 15% over a few years, with 10% as a more usual upper limit, in severely affected flocks before their dispersal. Most flocks of an affected inter-breeding population are free of the clinical disease in most years, with 1–2% of cases, or less, in occasional years.

(ii) The introduction of affected sheep, or of sheep subsequently developing clinical illness, into a flock free from clinical scrapie for some years does not necessarily lead to the occurrence of scrapie in the recipient flock, and in many flocks with controlled breeding practices never does. The introduction of stud rams subsequently developing the clinical disease into such flocks is not followed by any scrapie in the ewes mated by such rams, nor in their F_1 generation progeny.

(iii) The introduction of unaffected sheep, ewes or rams, from a flock of known scrapie-free background into a severely affected flock is not followed by clinical disease in the introduced sheep nor in their F_1 progeny.

(iv) The disease is more frequent among the progeny of affected parents or with close relatives affected, but the precise pattern varies greatly from flock to flock. The progeny of a particular ram in one flock may be heavily affected, while the progeny of the same ram in another flock is unaffected.

(v) A ewe affected during a gestation, or subsequently, tends to have affected progeny, but often has no affected progeny.

(vi) Progeny of an affected or of an unaffected dam born at different stages of their reproductive lives may be affected or unaffected irrespective of the dam's parity, of the progeny's birth order or twinning with an affected sib, or of clinical illness in the dam during the gestation, parturition or suckling of the progeny.

(vii) Both sexes are probably affected equally, although the dispersal by sale and slaughter of young males reduces the number under observation. Affected rams tend to show the characteristic rubbing and nibbling much less conspicuously than females, and their inanition may not be recognised as scrapie. Castration does not accelerate or

delay the expected development of the disease, nor does early pregnancy in the female.

(viii) Either parent of an affected animal may develop the clinical
disease after the offspring's death from scrapie.

This complex epidemiological pattern is difficult to reconcile with
any simple infection by a pathogen readily communicable to susceptible
sheep, but there are occasions, when limited information only is
available, which can appear to support an infectious origin. Thus
some observations have been interpreted as evidence for widespread
maternal transmission of an infectious pathogen from dam to offspring
(237, 238, 98). My own somewhat longer series of observations renders
these propositions unlikely.

Though I cannot deny the *possibility* of infection occurring naturally
in the field, my observations have led me to the conclusion that such
infection, if it occurs, is uncommon. Putting it at the lowest, none of
my observations seem to call for an explanation in terms of infection.
On the other hand, the hypothesis that natural scrapie is due to the
action of a single autosomal recessive gene fits the observed facts so
well (226–229) that it is reasonable to place the burden of disproof
upon those who reject the genetic hypothesis. In what follows, I shall
assume that the theory is valid, and speak in terms of dominant and
recessive genes and alleles.

II. The Three Scrapie Genotypes

The hypothesis of a recessive gene postulates that sheep will be of
three genotypes: (1) the double recessive, designated ss, which develops
the disease. For the sake of brevity, such an animal is termed 'black';
(2) the double dominant SS, which is unaffected, termed 'white'; and
(3) the heterozygous carrier Ss, which is also unaffected, termed
'grey'. Matings of 'black' × 'black' produce all 'black' (i.e. affected)
progeny; with 'black' × 'grey' half the progeny are affected; 'grey' ×
'grey' produce 1 affected progeny in 4, while with 'white' × 'black' or
'white' × 'grey' no progeny are affected. Data on these reciprocal
crosses have now been obtained in an experimental flock with a continuing high incidence of scrapie for most of the period, with some data

on matings with 'white' ewes from three other flocks. The results fit the predictions remarkably closely (see Appendix A).

If the recessive gene hypothesis is valid, then it should be possible to control scrapie in an affected interbreeding population, since no progeny of a 'white' ram should develop scrapie. In fact, none of them do develop the disease; and by the judicious introduction of the progeny of 'white' rams, the number of flock ewes carrying the allele s, i.e. the scrapie-allele frequency of the flock ewes, may be brought to a level where further cases of scrapie do not occur, even when the progeny of rams developing scrapie are retained. No selective removal of affected animals is necessary. In our breeding experiments, progeny with one or both parents affected have been retained as required for the test-mating programme.

As yet there is no laboratory test or simple way of knowing to which of the three genotypes any sheep belongs. Some 'blacks' with both parents affected may be predicted from their parents' records, but 'blacks' derived from a mating with a 'grey' parent cannot be identified by phenotype inspection, nor can the 'white' be recognised clinically. Such identification requires evaluation of progeny from special test matings. Since the practical control and possible elimination of scrapie depends on the identification and use of 'white' rams among those selected by breeders as acceptable to them as stud sires, this protracted test-mating procedure has proved worth while in practice. 'White' scrapie-allele-free animals can be identified by mating them with 'black' and 'grey' spouses to provide 10 progeny by a 'black' or 22 by a 'grey', living and well until at least $4\frac{1}{2}$ years old or later. Under present farm facilities this test-mating procedure can only be applied to rams: 105 rams have been put on test, of which 18 have been proven ($P <$ 0·01) not to carry the scrapie-allele. The rams are termed 'proven white'.*

---

*The argument here could be expressed as follows: assuming that the manifestation rate is 0·76 at age $4\frac{1}{2}$ (Appendix A, p. 161) the probability that an animal born to a 'black' ewe and a 'grey' or 'black' ram would be affected by age $4\frac{1}{2}$ is *at least* 0·5 × 0·76, i. e. 0·38, and the probability of its *not* being affected by this age is *at most* 0·62. Thus if the ram is carrying the scrapie gene, the probability that none of the offspring from 10 matings with 'black' ewes will be affected is at most $0·52^{10}$, which is less than 0·01. The same applies to matings between 'black' rams and 'grey' or 'black' ewes (G.J.D.).

III. Building-up a Scrapie-free Flock

1223 progeny of 'proven white' rams, 1167 female and 56 male, have been born and reared in 15 flocks; 825 females and 12 males have been born and observed to $4\frac{1}{2}$ years old. No case of scrapie has occurred in this progeny, although all 15 flocks have had cases of scrapie at some time during the observation period 1958–1978 (Appendix A).

A 'proven white' ram has been used widely in the 15–19 flocks in the breeder-group, and any suitable sons kept for study. The first 'proven white' ram was identified in 1962. Since then scrapie in the group flocks has declined from 3.5% in 1964/65 to 0.2% in 1975/76 and 1976/77, with no cases in home-bred sheep since the spring of 1977; i.e. the disease has virtually disappeared.

In one flock with an initial annual flock attack rate of 15%, and with over 50% of females in some birth-cohorts dying of scrapie for 5 successive years, 'proven white' rams have been used in part of the flock to control scrapie and develop a 'white' female section while breeding for scrapie in another portion. The 'white' section has been based on three unrelated 'proven white' rams used to provide consecutive 'white' male top-crosses on the paternal side of the pedigree without interruption, each generation being designated F_1W–F_5W etc. By the F_8–F_9W the frequency of the scrapie allele should be approaching 1 in 1000. Animals of the fifth to seventh 'white' generations are at present under observation, and are free of clinical scrapie, though not free of the scrapie allele.

It is of course characteristic of a recessive gene that it may remain latent over several generations. This seems to have occurred on at least two occasions in our recent experience. In the first, an F_5W ram, born in 1971, has been test-mated and shown by a daughter affected in 1976 to be a 'grey'. Its detailed extended pedigree shows no recent forebear as affected, or having affected progeny; but its maternal great-great grandmother, born in 1956, was affected in 1959. From this affected forebear the allele will have passed silently, without giving rise to clinical illness, through five generations over 17 years. In the second example another self-contained flock, which has not had a case of scrapie in home-bred stock for at least 23 years and probably much longer, accepted an introduction in 1973 from another flock, scrapie-

affected, of 10 young females sired by an affected ram. Two of these
ewes subsequently developed scrapie, but no case had occurred in the
rest of the home-bred flock until 1978, when two granddaughters of the
introduced ewes (which were themselves unaffected), sired by an
unaffected purchased ram which was being test-mated, became affected.
In this instance the clinical disease has missed one generation only, and
has not appeared outside the small group of females known to be
carrying the scrapie allele.

Probable Flock Allele Frequencies and 'Epidemics'

Where scrapie has appeared as an 'epidemic' in a flock of an inter-
breeding community, there is often evidence suggesting a gradual
increase in the scrapie-allele frequency among the flock females,
attributable to retention, over a decade or so, of the daughters of
unsuspected 'grey' rams, or of 'black' rams lost before manifestation of
the disease. The use of highly desirable, but allele-carrying, rams could
in theory result in progeny groups showing scrapie attack-rates of
around 50%, and could be responsible for the 'sudden' appearance of
flock incidences of 10–15% (227, 228, 106).

The decline and virtual disappearance of scrapie in Suffolk flocks of
our breeder-group between 1962 and 1978 may be attributed to a
reduction of the frequency of the natural scrapie allele in the flocks.
Some flocks have probably had a very low frequency throughout; some
have had a varying frequency, often high in certain families, while in
a few an extraordinarily high frequency has been accompanied by
'epidemics' so devastating economically that only the most careful
flock management has allowed the flocks to survive. The different
experiences are related to the fine details of the types of stud rams
which individual flockmasters and their shepherds tend to select over
the years. It appears that the scrapie allele is associated with certain
phenotype characteristics of muscular and carcase development in
adolescence which are sought for by many breeders (226, 284), and
that there is a selection pressure for the scrapie allele. The situation
in any flock represents a balanced genetic polymorphism, with the
level of allele frequency determined by the sire-selection preferences
(226).

IV. Other Postulated Genes

This autosomal recessive gene, which appears to control the clinical manifestation of natural scrapie, must be clearly distinguished from other proposed genes which have been thought to influence the response of sheep to the transmissible agent, namely (1) a dominant gene associated with the 'artificial' post-inoculation disease (212) and (2) a recessive gene postulated by Dickinson (98) in connection with his proposal for the vertical transmission of an infective agent through the dam. Any assessment of the reality of these two proposed genes, of their relation to the recessive gene with which we have been dealing, and of their relevance to the natural disease, must await a great deal more detailed information.

It suffices to say that the natural history of spontaneous scrapie may be accounted for very precisely by postulating a trait or entity carried equally by the cells of both sexes, which in the single state does not produce ill-health but when in the double state, half obtained from each parent, leads to a premature decay of certain brain cells with the early disability and death characteristic of scrapie. The behaviour of this entity during reproduction is compatible with an autosomal Mendelian recessive gene. The transmissible agent is probably generated during the metabolic disturbance responsible for the decay of the brain cells (226, 231) and plays a subsidiary role in the disease process.

V. Dissemination of the Disease by the Allele Rather than the Transmissible Spongiform Encephalopathic Agent (TSEPA)

As the recessive gene hypothesis offers such an accurate quantitative explanation for the differing patterns of occurrence of the clinical disease, and has allowed precise predictions of its occurrence, reduction and disappearance, now confirmed by direct observation in many flocks, there seems good reason to accept that the gene is the main means for the dissemination of the disease. A small proportion of cases may possibly be due to inadvertent contamination of non-scrapie genotypes with the transmissible agent, for instance by the ingestion of placental remains containing the agent (237, 238), although observation of ewes' *post partum* behaviour suggests that this is likely to be a rather rare occurrence in most flocks. The possibility of transovarian and transplacental transfer of the agent has been suggested (98); if this occurs, one has to explain why any lamb sired by a 'white' ram remains unaffected when its dam is actually affected clinically, and why the birth order of affected progeny among the total completed families of unaffected 'grey' or affected 'black' dams is so varied.

It is, of course, possible to build an explanatory hypothesis on the basis that the recessive gene confers absolute susceptibility to the transmissible spongiform encephalopathic agent (TSEPA) of scrapie, which must be of widespread or even ubiquitous distribution, since any double recessive genotype would have to be exposed to the agent at the right time during its early life when its susceptibility was total. True, the agent is very stable and not easily destroyed in the laboratory, but we have no reason to believe that any such agent has been widespread in Western Europe since the middle of the 18th century.

Other data offer little support for such an hypothesis. Assays for the presence of a TSEPA in normal sheep of non-scrapie genotype from severely affected flocks have failed to demonstrate any TSEPA, while in affected sheep of the same birth-cohort the TSEPA is present. TSEPA has been found in clinically normal sheep, but only in those showing early neuropathological lesions. The nature and distribution of the primary lesions are remarkably constant and very localised, with a sequence of decay reminiscent of genetically determined abiotrophic degenerations of other specialised nerve cells, such as the rod cells of the retina (223). It is possible that the TSEPA may initially be restricted to those special nerve cells; but no data on the distribution of the scrapie TSEPA among the various cell components of the brain have been published.

Until reliable methods are available for determining the distribution of TSEPA in the field, for indicating the present or past exposure or harbouring of a TSEPA in the clinically normal animal and in the scrapie-prone genotype, assessment of the role of a communicable infective agent in the dissemination of scrapie is impossible. Attempts to draw analogies from the large body of observations on the artificial disease and on the biochemical and cellular nature of the agents derived from mouse and other tissues, while of great scientific interest, are of little immediate relevance to the practical problem of controlling or eliminating scrapie in sheep.

Chapter 7

Pathology of Natural Scrapie

In 1950 our knowledge of the pathology of natural scrapie rested very largely on French studies, notably that of Bertrand *et al.* (33) augmented by the English reports of Brownlee (53) and Holman and Pattison (151). The principal, and by many considered diagnostic, feature was the presence of large vacuoles, apparently empty, in the cytoplasm of neurones in the mid-brain, pons and medulla, and in the spinal cord, especially in the ventral and lateral horns. Bertrand *et al.* also stressed astrocytic proliferation and gliosis in these sites, and particularly in the cerebellum and in the supraoptic region of the hypothalamus. Others, however, failed to find lesions in some of these sites, and indeed their occurrence is variable. Thus many workers relied entirely on the presence or absence of vacuolated neurones in the medulla (151). For a summary of earlier investigations, see Table II of Beck *et al.* (23).

The presence of vacuolated neurones in sections from an intravitally fixed brain, using the technique developed for the study of retinal degenerations of the dog (223), established for us that the vacuoles were not due to post-mortem change, and might be accepted provisionally as a basis for diagnosis. However, it was difficult to reconcile the observed clinical signs with such a limited morphological base, especially as between 1954 and 1956 signs were frequently observed indicating skeletal muscle involvement, such as inability to hold the head in the horizontal plane except for short periods, associated with pallor of the midline dorsal craniocervical neck muscles, asymmetry of hind-limb muscles, and a low tolerance to muscular exercise. A thorough autopsy examination of the skeletal musculature, rarely examined in routine veterinary work, revealed a surprising number of affected animals with skeletal muscle changes of myositic type, not found in the control animals (43). This led to the concept of scrapie as a primary myopathic disease (224). Vacuolated neurones were present in the brains of these animals with myopathic lesions, but it was clear that these were inadequate to explain the observed clinical signs of disturbed control of body temperature, cardiovascular responses and water intake, while the myopathic signs were suggestive of a general metabolic disturbance, possibly related to the enlarged adrenal glands which we had observed, rather than to a purely muscular disorder. A myopathy identical histologically to that in our scrapie sheep had been produced in rabbits by administration of cortisone (110), a myopathy which we found could also be produced in lambs. Since then there have been many reports of myopathy in human patients with endocrine disorders (3).

Previous experience of a syndrome somewhat similar to the forms of scrapie we were observing had occurred in the University of Sydney in an Australian Army Remount Horse in 1943. After considerable observation, a clinical diagnosis of hypothalamic malfunction was made, and confirmed at autopsy, when a very large *Echinococcus* cyst was found, causing extensive compression atrophy of the ventral hypothalamic region. In the late 1950's experimental studies on the effects of interruption of the pituitary stalk and of removal of the pituitary gland (hypophysectomy) were being conducted by Daniel and Prichard (89, 90). Observation of their operated animals suggested that at least some of the homoeostatic malfunctions noted in scrapie might be related to the region of the hypothalamus and pituitary gland. They were persuaded to operate on some Welsh mountain pregnant sheep; and it was the study of the long-term survivors of these experiments, many completing their pregnancies, which provided an essential clue. It was found that a substantial loss of neurones in the supra-optic and paraventricular nuclei was occurring in scrapie, similar to that seen in the postoperative experimental animals (20, 22) and in human brains following traumatic and other transections of the pituitary stalk (91). Without sections of this control surgical material, seen alongside the scrapie brain sections, recognition of small degrees of neuronal loss would have been extremely difficult to establish, and the significance of an occasional vacuolated neurone open to question.

It must be stressed that even today there is no comprehensive detailed account of the normal macro- or micro-anatomy of the nervous system of the sheep, although studies of the lower brain stem have been made (220, 151). Only a short stereotaxic study of one large-brained breed, the French Pré-Alpes du Sud, is available (251). Without an accepted corpus of information such as is available for man and for certain laboratory animals such as the rat, mouse and cat, it is extraordinarily difficult to recognise small changes in the neuronal populations of the various parts of the brain and spinal cord. The observations of Beck and Daniel were of the greatest importance, and have led to the recognition of widespread specific neuronal loss in scrapie, confined initially to certain anatomical-physiological 'systems' (or pathways), namely the afferent input to the cerebellum from the pons and inferior olives, and the brain-pituitary pathway via the pituitary stalk.

From this work, many clinical features of scrapie have acquired a new and more satisfactory explanation, based on physiological malfunctions due to specific neuronal degenerations confined to certain 'systems', notably the hypothalamo-neurohypophysial and the olivo-ponto-cerebellar. This discovery is almost entirely due to the work of Professor P. M. Daniel and Mrs. E. Beck.

From this base the concept has emerged (23, 88a, 227) that scrapie may be considered as a spontaneous degeneration of specific neural structures, akin to the *primary neuronal degenerations* of man (214), but belonging to a special group characterised by a substance or agent capable of initiating a similar degeneration when artificially introduced into an experimental animal. Recent electron microscopic and immuno-histochemical studies, described in the next chapter, provide an insight into the possible metabolic characteristics of the nerve cells which are at special risk, and of the fundamental malfunction which may underlie their shortened life-span.

I. Autopsy Techniques and General Observations

Animals have, whenever practicable, been examined alive prior to autopsy; then anaesthetised, without excitement, by intrajugular injections of 10–40 ml of Nembutal (60 mg/ml, w/v of pentobarbitone sodium, Abbott); and killed by exsanguination following a bold incision in the neck immediately caudal to the thyroid glands, cleanly severing both carotid arteries and both jugular veins. Such exsanguination makes examination of the nervous system much cleaner and more precise; it also tends to reduce the weight of individual abdominal viscera, especially the spleen, where barbiturate anaesthesia is liable to produce marked vasodilatation. Only rarely and under special circumstances has a defunct animal been examined, and then usually within 12 hours of death. Special care has been taken to avoid the complications of protracted terminal illness, such as dehydration and inanition associated with failure to ingest food, and the self-inflicted muscle and skin injuries which may result from recumbency. Normally the brain is removed first. The meninges are cut to reduce traction on the pial vessels as far as possible. The brain and upper cervical cord down to C2 are carefully removed, and the cord then severed 2·5 cm below the caudal end of the fourth ventricle. The brain is weighed, and immersed in at least 6 volumes of 10% neutral formol saline in a wide-mouthed jar. Occasionally other fixatives have been used. If material is required for use with the Holzer stain, the brain has been left *in situ*, or covered on the bench, for 8 hours before being placed in fixative. In a number of animals the whole spinal cord was exposed, and the thoracic and cervical nerve roots and their dorsal root ganglia were dissected out.

The general autopsy examination normally shows no conspicuous changes upon routine inspection. The subcutaneous tissues are often rather dry, and separate with an unusual crackling sound, recognised by some experienced slaughterhouse attendants as indicative of scrapie. Some organs, such as the endocrine glands, may show departures from normal weights, but in general the brain, heart, liver, kidneys, spleen and reproductive organs are within normal limits. Under these circumstances, when we are looking for small changes of a physiological rather than of a gross pathological disturbance, and in the absence of any published table of sheep organ weights, it is important to establish the normal weight ranges for animals of the same breed, age, weight, reproductive stage and season of the year.

The brain weights, which vary considerably between breeds and with body weight, were within the normal range; some showed a modest increase in size of the ventricular system. The weight of the pituitary gland varied considerably, the anterior lobe being often enlarged, for reasons which could not be ascertained. The normal range was 1·2–1·8 g, but occasionally glands weighing 2·5–3·0 g occurred in apparently normal sheep. The lower infundibular stem was often small and grey in colour, and frequently deeply enveloped by anterior lobe tissue, while the posterior lobe was somewhat shrunken, and greyer than usual. The pars tuberalis and the area of the median eminence were more prominent and pinker in colour.

The thyroid glands were small and translucent, usually less than a total 2·5–3·0 g, while the adrenal glands exceeded 4·0 and sometimes 6·0 g. The thyroid weight : adrenal weight ratio was greater than 1·0 in the normal controls, but always less than 1·0 in animals with scrapie. The liver and kidneys were of normal appearance, but the liver tended to be somewhat enlarged though of normal consistency and texture. No specific abnormalities of the reproductive tract were encountered, in keeping with the fact that pregnancy, parturition and suckling proceed normally in affected ewes until stage 3, or until behavioural anomalies lead to incessant movements. The heart was apparently normal, but often not closely examined. The spleen and lymph nodes were of normal size and consistency. The skeletal musculature usually showed loss of muscle mass, and from time to time massive areas of muscle pallor, such as were observed for several years in the 1950's (43) (Fig. 7.9) have been seen.

During the early years of the investigation, for some 800 animals, fairly complete autopsies, each taking half a day, with the dissection and weighing of all main organs (except the spinal cord, skeletal muscles and peripheral nerves), and with only a partial inspection of the skeletal muscles, were carried out. After the early 1960's, when the importance of the system degenerations of the brain and spinal cord became clear, and with the large numbers of animals required for our genetic studies, autopsy was restricted to the removal of the whole brain and endocrines, and rapid inspection of the abdominal viscera, primarily to establish that no other obvious cause of the illness or abnormality was present. On the other hand, doubtful and atypical cases were submitted to an appropriately detailed scrutiny at autopsy.

After fixation for 4 weeks, and with three changes of fixative, blocks of tissue were taken for routine light microscopy. Corresponding blocks from normal and from scrapie-affected animals were labelled in such a way that the person examining them did not know the animal's clinical status. Batches of material from 12 post mortems were sent to the laboratory, each batch including tissues from at least two normal and at least two clinically affected animals.

The initial study (23) was carried out on 34 affected animals (29 female and 5 male) aged 1½ to 5½ years, with an average of 3 years, and an average duration of the illness of 5·4 months. The controls were 12 clinically normal animals, of comparable age and parity, drawn from flocks not known to have harboured scrapie.

For details of the histological methods employed reference may be made to Beck *et al.* (23). The following were examined in every case: pituitary stalk, hypothalamus, thalamus, corpus striatum, subthalamus, different areas of cerebral cortex, four levels of brainstem, cerebellar hemisphere including dentate nucleus, vermis and roof nuclei, and upper cervical cord segments. The pituitary glands were examined in 29 cases, and muscles from 25 cases.

II. Histological Findings

In earlier reports on the pathology of the central nervous system in scrapie the emphasis has been mostly on the presence of vacuolated neurones in the brainstem and spinal cord of affected animals; appreciable lesions in the white matter have not been described. The only comprehensive pathological investigation was that by Bertrand *et al.* (33) on a series of 20 animals; but these authors gave no detailed account of the clinical illness in their animals. Indeed it has been and remains common practice to base the diagnosis of scrapie on the occurrence of vacuolated neurones, usually in the medulla only, in the lateral cuneate nucleus, and in the dorsal nucleus of the vagus, irrespective of the neurological signs presented by the animal in life.

These vacuoles, 5–10 μm in diameter, are seen in the cytoplasm of neurones. They are apparently empty, and sharply delimited by a membrane, the presence of which has been confirmed (p. 132) by electron microscopy. The nucleus is displaced to the periphery of the cell body, and there is marked reduction of Nissl substance (Fig. 7.1). Vacuolated neurones are said to occur in clinically normal sheep (151, 338, 339, 218), but in significantly smaller numbers (332). As the 'normal' animals used in these studies were of unknown ancestry, the presence of preclinical scrapie cannot be excluded. In the 12 normal control animals, of known pedigree and health records over three generations, used by Beck *et al.* (23), vacuolated neurones were absent.

This study was the first attempt to correlate the clinical signs with the pathological lesions, and differs from previous reports by demonstrating a spontaneous degeneration in two major anatomico-physiological systems—the cerebellar system and the hypothalamo-neuro-

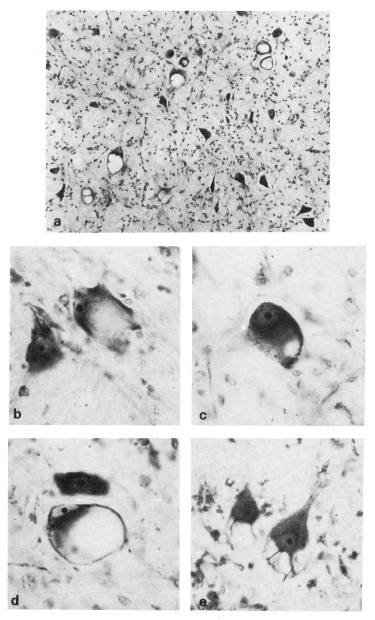

Fig. 7.1. *Neuronal vacuolation:* (a) *low-power view of lateral cuneate nucleus in sheep with scrapie. Nissl, magnification* × *82;* (b) *cell from lateral cuneate nucleus, showing classical central chromatolysis (axonal reaction), with nucleus displaced to periphery. Nissl, magnification* × *450;* (c) *small, and* (d) *large vacuoles within chromatolytic cells. Nissl, magnification* × *450;* (e) *two Purkinje cells, showing 'primary' vacuolation. Note the normal central position of the nucleus. Nissl, magnification* × *450. (Figs 7.1–4 and 7.6–8 are taken from Beck et al. (1964) by courtesy of Oxford University Press.)*

hypophysial (HNH) system—and by suggesting that neuronal vacuolation can be interpreted as part of the 'system' degeneration. The following account is based on this paper, which should be consulted for further details.

In 21 out of the 34 cases examined both the cerebellar and the HNH systems were affected, though to a varying extent. In six cases the cerebellar system alone was involved, the HNH system being essentially intact, while in another seven cases the HNH system was mainly affected. This difference in the distribution of the pathological lesions was paralleled by clinical differences; whereas in some cases cerebellar signs and metabolic disturbances were equally marked, in other cases one or the other predominated.

While a close correlation existed in all cases between the severity of clinical cerebellar signs and of pathological changes within the cerebellar system, the correlation between the metabolic disturbances and the pathological changes within the hypothalamus was not as close. Neither the age of the animal at onset of the disease nor the duration of the illness seemed to be related to the severity of the pathological changes (23, Table I). The number of rams examined was small, and it has thus not been possible to decide whether sex influences the severity of the changes.

The cerebellar system

In 29 animals clinical signs of cerebellar disturbance such as ataxia and tremor were seen during life, and in all of these the degree of the cerebellar signs correlated well with the severity of the lesions found in the cerebellar system. The remaining five animals had no obvious cerebellar signs, and in all of them the cerebellar system showed only very slight changes.

The pathological changes were found bilaterally in the afferent parts of the cerebellar system, while efferent cerebellar pathways remained comparatively unaffected. The loss of nerve cells was particularly marked in the granular layer, and in the Purkinje cells of the flocculonodular lobe, i.e. the palaeocerebellum was most markedly affected. If the cases were classified according to the clinical cerebellar signs, then all animals in which these were severe or moderate showed degeneration as follows:

1. Severe to moderate loss of nerve cells in both the cerebellar cortex and in the pontine and papilioform nuclei, with corresponding reduction of nerve fibres (Fig. 7.2).

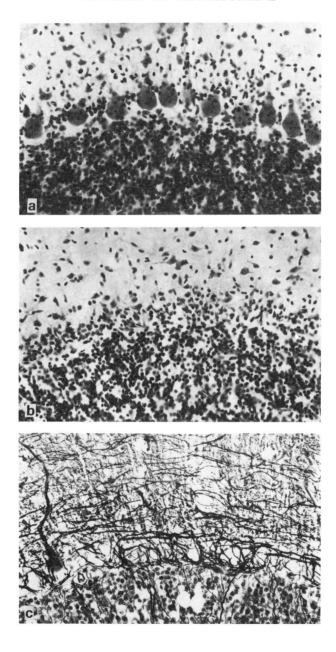

Fig. 7.2. *Cerebellar cortex.* (*a*) *Normal control.* (*b*) *Case showing complete loss of Purkinje cells and rarefaction of the granular layer; Bergmann glial cells are only slightly increased in number.* (*c*) *Case showing line of empty baskets.* (*a, b*) *Nissl.* (*c*) *Marsland-Glees. Magnification* × *190.*

2. Myelin breakdown and dense to moderate fibrous gliosis of grey and white matter in these regions (Fig. 7.3).

3. Striking nerve fibre degeneration (degenerating mossy endings, retraction bulbs and torpedoes) in the cerebellar cortex of the less severe cases (Fig. 7.4).

4. Variable changes in the inferior olives, and severe to moderate neuronal vacuolation in other brainstem nuclei.

In some cases in which the cerebellar system was only moderately affected, fibrous gliosis was entirely confined to the nodulus, a part of the cerebellum concerned primarily with vestibular function. The dentate and roof nuclei were well preserved in most cases.

All animals free from clinical cerebellar signs showed the following changes in the cerebellum: vacuolation of some Purkinje cells, some degeneration of mossy endings, proliferation of Bergmann glial cells (but no fibre formation) and oedematous detachment of the molecular and Purkinje cell layers from the underlying granular layer. Occasional vacuolation was found in the brainstem nuclei. Even where degeneration of mossy endings in the cerebellar cortex was severe, clinical signs of ataxia and tremor were not always present. The findings in three animals not included in this study are of interest in this connection (19). These animals which, because of their known hereditary background, would have been expected to develop scrapie at a later stage (107) were at the time of autopsy free from clinical signs. All three of them showed, as the only pathological finding, innumerable degenerating mossy endings throughout their cerebellar hemispheres (Fig. 7.5). This was probably the first indication of cerebellar system degeneration.

It was observed that although large numbers of degenerating mossy terminals in the cerebellum seem to be compatible with normal coordination of movement, large numbers of degenerating *boutons terminaux* around the apparently normal neurones of the pontine and papilioform nuclei were associated with a marked degree of ataxia.

In all cases with very severe degeneration of the cerebellum the pontine and papilioform nuclei as well as the middle cerebellar peduncles were severely affected. There were, however, some cases in which cerebellar changes were more severe than those found in the pons and some others where pontine changes predominated. Similarly, degenerating mossy endings were found in the cerebellum of one case in which the pons was normal, while degenerating *boutons terminaux*

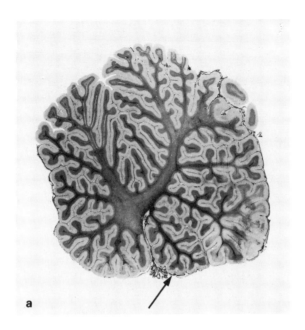

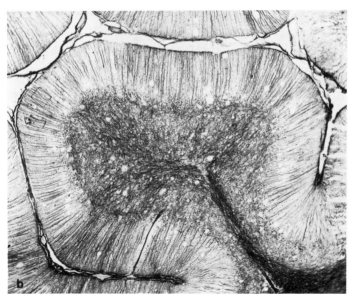

Fig. 7.3. *Gliosis in cerebellum. (a) Sagittal section through the vermis. Holzer, magnification × 2.5. (b) Higher magnification of one folium. Holzer, magnification × 50. There is severe fibrous gliosis in cortex and white matter. The nodulus (arrow) is most severely affected.*

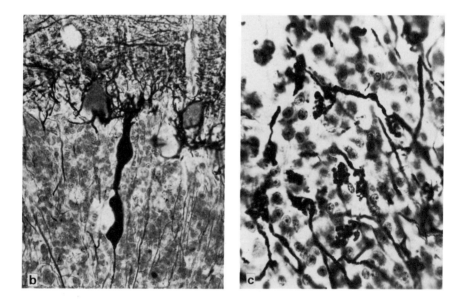

Fig. 7.4. *Examples of different types of nerve fibre degeneration within the cerebellar cortex.*
(a) Diffuse break-up and swelling of nerve fibres; (b) torpedo-like swelling of Purkinje cell axon;
(c) degenerating mossy terminals. Gros-Bielschowsky; magnifications: (a, b) × 200; (c) × 600.

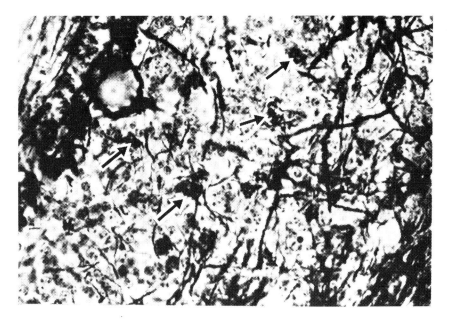

Fig. 7.5. *Cerebellum from a ewe, killed at 3¼ years, clinically normal, but progeny developed scrapie. The only pathological change found in this case was a widespread degeneration of mossy endings (arrows) within the granular layer of the cerebellum. In a further five cases of potential scrapie degenerated mossy endings were also found. Gros-Bielschowsky silver preparation, magnification × 480. (Figure taken from Beck (1969) (Fig. 36) by courtesy of Blackwell Scientific Publications.)*

around the neurones of the pontine nuclei was the only significant finding in another case; finally, in two further cases there were degenerating mossy endings in the cerebellum as well as degenerating *boutons terminaux* in the pons. It has thus not been possible to decide where the disease process starts, whether in the cerebellum or in the pons, and in fact, judging from the cases showing only acute degeneration of nerve terminals, the site may well vary from one animal to another. In all cases of advanced degeneration the changes become widespread, so that evidence as to the starting point is obscured.

The degree of degeneration in the inferior olives was variable, and did not correlate as closely with cerebellar changes as did that in the pons. Thus, of two animals, both of which had severe cerebellar lesions, the inferior olives were normal in one but showed virtually complete nerve cell loss, accompanied by fibrous gliosis, in the other. The most common change was a dense fibrous gliosis (Fig. 7.6), together with an

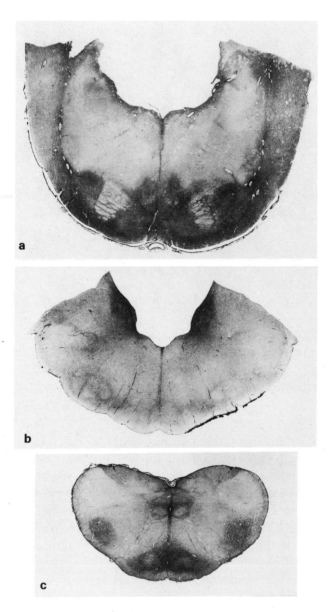

Fig. 7.6. *Transverse sections through (a) the pons at the level of the main trigeminal nuclei, (b) lower pons, including the medial vestibular nuclei, (c) the lower medulla, stained by the Holzer method for glial fibres; all magnifications × 4. Note fibrous gliosis in pontine and papilioform nuclei and the lateral parts of the middle cerebellar peduncles, with sparing of the corticospinal tracts, in (a); of the medial vestibular nuclei in (b); and of the inferior olives and lateral reticular nuclei in (c). Compare with Fig. 7.7.*

extreme neuronal atrophy and hyperchromasia, a phenomenon interpreted by Brodal (48) as being due to retrograde degeneration, and
frequently observed by that author following experimental cerebellar
ablation in rabbits and mice.

Degeneration in other nuclei of the brainstem and spinal cord further
stressed the systematic character of the lesions, since it was always
bilateral and essentially confined to structures belonging to the cerebellar system, i.e. the medial and lateral vestibular nuclei, the external
(accessory) cuneate nuclei, the nuclei of the brainstem reticular formation, the red nuclei and the neurones of Clarke's columns (see Fig.
7.7). Furthermore, the intensity of the changes roughly corresponded
with the severity of the cerebellar degeneration in the individual
animals.

Although the nerve cells of Clarke's columns showed changes in
six of the nine cases in which lower levels of the spinal cord were
examined, the dorsal spinocerebellar tracts appeared normal in all
cases except one, in which both dorsal spinocerebellar tracts showed
marked degeneration in the cervical region.

There is thus amongst the 29 ataxic animals studied by Beck *et al.*
a variety of patterns in which the cerebellar system may be affected.
Firstly, and most commonly, as an *olivo-ponto-cerebellar degeneration*;
secondly, and next in frequency, as a *ponto-cerebellar degeneration*; and
thirdly, in one case only, as a *spino-ponto-cerebellar degeneration*. Since the
sheep with spinal involvement came from the same flock as several of
the animals showing the other two types of degeneration, and since
each flock may in itself be regarded as one large family, it became
obvious that scrapie, as a system degeneration, may take different
forms within the same family. That there may be even greater variations between different 'families', i.e. breeds, is emphasised by the
findings of Besnoit and Morel (34) and Bertrand *et al.* (33) who,
working in France, necessarily investigated different 'families'. It is
therefore not surprising that their results included some additional
features, such as peripheral neuritis and vacuolation of anterior horn
cells. Thus their cases may represent still other variants of the disease.

Most investigators have emphasised the presence of vacuolated
neurones in the brains of sheep suffering from scrapie, and although
reports have varied regarding their overall distribution, there has
nevertheless been general agreement, confirmed by Beck *et al.* (23), on
their predilection for certain brain stem nuclei, which are always

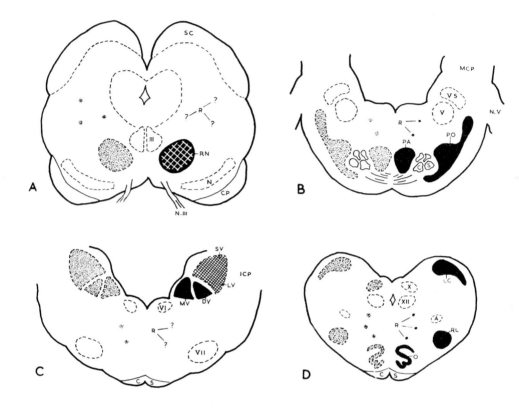

Fig. 7.7. *Diagram to show the nuclei in the brain-stem of the sheep (at levels which have been investigated in every case). (A) mid-brain; (B) mid-pons; (C) lower pons (with a composite representation of the vestibular nuclear group); (D) mid-medulla. The principal nuclei with known cerebellar connexions are shown on the right; nuclei with efferent connexions to the cerebellum in solid black, nuclei with afferent connexions from the cerebellum cross-hatched. White cross-hatching on black in the red nucleus indicates a two-way connexion. (For detailed connexions of vestibular nuclei and brain-stem reticular formation see Brodal et al. (1962) and Brodal (1957).) On the left, stippled, are the nuclei which usually contained appreciable numbers of vacuolated neurones and/or showed other signs of degeneration. Note that only those nuclei which have cerebellar connexions showed signs of degeneration (with the exception of the dorsal motor nucleus of the vagus nerve). Abbreviations: A, nucleus ambiguus; CP, cerebral peduncle; CS, corticospinal tract; DV, spinal vestibular nucleus; ICP, inferior cerebellar peduncle; LC, lateral cuneate nucleus; LV, lateral vestibular nucleus; MCP, middle cerebellar peduncle; MV, medial vestibular nucleus; N, substantia nigra; N III, NV, cranial nerves III and V; O, inferior olive; PA, papilioform nucleus (component of pontine nuclei); PO, pontine nuclei; R, brain-stem reticular formation; RL, lateral reticular nucleus; RN, red nucleus; SC, superior colliculus; SV, superior vestibular nucleus; III, V, VI, VII, X, XII, motor nuclei of cranial nerves; VS, sensory nucleus of V. The terminology used is that adopted by Palmer (1958).*

bilaterally affected. Despite this, the aetiology of these vacuoles has remained obscure, and they have generally been regarded as a primary response of the neurone to the disease process. In the light of the present findings, however, an alternative interpretation seems possible. With the exception of the dorsal nucleus of the vagus, which is now known to have connections with the paraventricular nucleus (290), only those nuclei that are known to project to the cerebellum have contained vacuolated neurones in large numbers, whilst adjacent nuclei, which do not project to the cerebellum, have appeared essentially normal (see Fig. 7.7). Moreover the number of affected nerve cells in the brainstem nuclei has been roughly proportional to the degree of the cerebellar degeneration found in each individual case. This striking finding has been interpreted as follows.

Most afferent fibres entering the cerebellum terminate as mossy fibres within the granular layer (292, 179, 187). Degenerating mossy terminals are a common feature in cases with early generalised cerebellar changes. Degeneration of its mossy terminal may cause the nerve cell of origin, i.e. the neurone in the affected brainstem nucleus, to react with central chromatolysis. Such cells, with eccentric nucleus, a thin rim of Nissl substance and a clear central cytoplasm, have been noted by several investigators (33, 333, 334), and were also observed in our own material side by side with vacuolated neurones within the same brainstem nucleus (Fig. 7.1b, c, d). That vacuolation may develop in a cell undergoing retrograde change was observed by Holmes and Stewart (152) in the inferior olives of human patients with acute cerebellar lesions, but not with long standing ones. Likewise Jacob (162) illustrated both central chromatolysis and vacuolation in adjacent cells in the anterior horn of a patient with compression of the cauda equina. Central chromatolysis is now thought to indicate an increased protein synthesis within the cell, which in extreme cases may lead to vacuolation (149; 'signet ring cells'). The same interpretation would probably apply to vacuolated nerve cells in the red nucleus, although here vacuoles could also be due to transneuronal degeneration which, according to Jacob (162), may lead to the same change. The existence of a two-way connexion between red nucleus and dentate nucleus has been confirmed by Brodal and Gogstad (50), and the fact that in scrapie material vacuolated nerve cells were comparatively sparse in the red nucleus may be related to the relatively good preservation of the dentate nucleus in most of our cases. Vacuolated neurones in the brainstem and spinal cord of sheep with scrapie may therefore be interpreted as a sign of retrograde degeneration due to a rapidly developing cerebellar cortical lesion; the change may thus be described as 'retrograde' vacuolation.

In contrast to the 'retrograde' vacuolation seen in the brainstem nuclei, in which the neurones usually show eccentrically-placed nuclei and vacuoles which may be situated in any part of the cytoplasm, 'primary' vacuolation of Purkinje cells presents a different picture (Fig. 7.1e). It is usually multilocular, and is probably of different aetiology, being associated with oedema of the bodies of glial cells enveloping the base of the Purkinje cell (Beck, 1978, personal communication based on electron microscopic studies). Here the vacuoles are situated at the base of the cell, the nucleus remaining in the centre even when the vacuoles extend as far as the nuclear membrane; moreover, the occurrence of such cells frequently coincides with stretches of detachment of the molecular and Purkinje cell layers from the underlying granular layer. Primary neuronal vacuolation, far from being specific for scrapie, has been observed in many conditions where oedema is a prominent feature (219) and may possibly be caused by it. This is also suggested by the experimental production of vacuolated neurones in cats after an injection of distilled water into the common carotid artery (249).

The fact that myelin-stained sections of the brain and spinal cord appeared practically normal to the naked eye caused Bertrand *et al.* (33) to reject the idea of scrapie as a system degeneration. Nevertheless they were impressed, from the clinico-anatomical point of view, by the similarity of scrapie to a certain human system degeneration—a rapidly fatal form of amyotrophic lateral sclerosis. In our material, naked-eye myelin loss was usually not a striking feature, although myelin breakdown and accumulations of neutral fat were found within the cerebellar white matter and in the middle cerebellar peduncle in many cases. Most of this neutral fat lay free in the tissue; little was contained in phagocytes, and accumulations around blood vessels were rare. This unusual distribution is unlikely to be due to a disorder of any of the various types of glial cells, since the myelin was fully broken down, the periodic acid-Schiff (PAS) reaction was always negative, and oligodendroglial cells were present in normal numbers.

Hypothalamo-neurohypophysial (HNH) system

Clinical signs pointing, either directly or indirectly, to a disturbance of hypothalamic function have been described in sheep with scrapie for almost as long as the disease has been recognised. Emaciation, obesity and excessive thirst are some of the features given prominence in many reports; in addition, a craving of the animals for salt, intolerance of cold, and fluctuations in body temperature and heart rate have been observed. It is therefore surprising that no serious pathological study of the hypothalamus in this disease had been reported prior to Beck *et al.* (23).

While the correlation between clinical cerebellar signs and lesions in the cerebellar system was very good, that between metabolic and autonomic disturbances and degeneration in the HNH system of the individual sheep was less striking. Nor was the correlation improved when changes in other hypothalamic nuclei, or degeneration of a systematic character in other parts of the 'visceral brain', were taken into account. This lack of correlation however is not surprising, since the HNH system represents only one part of a complex regulatory system, whose connexions, interrelationships and functions are still not fully understood.

The supraoptic and paraventricular nuclei of the hypothalamus produce a neurophysin-containing neurosecretory substance which contains the antidiuretic hormone, and controls water metabolism, acting via the posterior lobe of the pituitary gland (see Parry and Livett, 231). Lesions within this HNH system lead to diabetes insipidus (see Daniel and Treip, 91). In scrapie degeneration of this system, though providing a pathological basis for the animals' excessive thirst, does not in itself explain other metabolic disturbances. However, there are a number of reports in the literature concerning human cases of diabetes insipidus (see Locke and Schally, 188), some of them hereditary, who showed in addition to the disturbed water balance various other abnormalities such as emaciation, obesity and fluctuations in blood pressure and body temperature. The metabolic abnormalities have been recorded in such cases even in the absence of any obvious lesion of the pituitary. The mode of interaction between different hypothalamic centres remains imperfectly understood. It seems probable, however, that degeneration of one major hypothalamic system could disturb the balance which exists between the hypothalamus, the pituitary gland and its target organs, and thus lead to a wide variety of metabolic and autonomic disturbances. Moreover, neurosecretory substances produced by the ventromedial hypothalamic nuclei, i.e. the hypophysiotrophic area, contain tripeptide hormones which regulate in part the activities of the anterior lobe of the pituitary, reaching the gland via the hypophysial portal system of blood vessels (136).

I know of no record of a disease in either man or animals in which spontaneous degeneration of the HNH system occurs in adult life. When this system was first studied in sheep with scrapie, a striking resemblance was noted to the retrograde changes seen in the same region in animals which had undergone the operation of hypophysectomy or pituitary stalk section (20). Both in sheep with scrapie and in sheep which had been operated upon there was a similar loss of neurones in the supraoptic (Fig. 7.8) and paraventricular nuclei, while many of the surviving neurones contained neurosecretory material. There was degeneration (and probably regeneration) of nerve fibres in the HNH tract, and excessive deposition of neurosecretory material, confirmed as containing neurophysin by immunohistochemistry (see Chapter 8), in the median eminence, in the pituitary stalk, and in the pars tuberalis (where we have not found it in normal animals of this species). Finally, as an early change, there was mesodermal proliferation in the territory of the two nuclei. Altogether the picture in the hypothalamus was so similar that in this one respect it was impossible to distinguish between a case of stalk section of some months' survival and a case of scrapie with well-marked HNH degeneration.

From this remarkable similarity it may tentatively be concluded that in scrapie degeneration of this system is retrograde in character and due to an initial axonal damage, although, so far, demonstration of degenerating terminals in the posterior lobe of the pituitary gland has not been achieved. That in scrapie there is also an attempt at regeneration (which is such a striking feature after pituitary stalk section) is suggested by the presence of nerve fibres and neurosecretory material amongst the glandular cells of the pars tuberalis, as well as by the apparently normal density of nerve fibres in the HNH tract, even in cases with maximal nerve cell loss in the supraoptic and paraventricular nuclei. This reaction occurs during the preclinical stage of the disease (see Chapter 8).

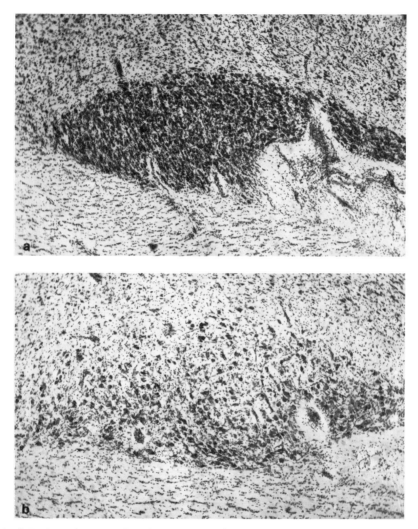

Fig. 7.8. *Coronal sections through supraoptic nucleus; (a) normal control; (b) from a case with severe loss of nerve cells. Nissl, magnification × 53.*

Degeneration in other hypothalamic nuclei was usually most conspicuous in those cases with severe lesions in the HNH system and less striking or absent when this system was less involved. Vacuolated neurones in the ventromedial hypothalamic nucleus and in the lateral hypothalamic area could be tentatively interpreted, in the same way as those in the brainstem nuclei, as being due to retrograde degeneration. Since, however, pathways in the hypothalamus are by no means as well mapped out as they are in the cerebellum, this question must remain open.

One case of Icelandic 'rida' (on present evidence, identical with scrapie), kindly sent by Dr. F. A. Pálsson, showed similar degeneration of the HNH and cerebellar systems. These changes have been reported by van Bogaert *et al.* (311), who also found lesions in the small suprachiasmatic nucleus.

The central autonomic system

Vacuolated neurones are frequent in the dorsal nucleus of the vagus, and in one breed have been used as a rapid method of confirming a clinical diagnosis. This nucleus was affected in 18 of Beck *et al.*'s 34 cases of a different breed; 14 showed vacuolation with fibrous gliosis, the rest showing either vacuolation or gliosis. Some loss of neurones has been observed in the intermediolateral columns of the thoracic and lumbosacral cord (321). The significance of such observations has hitherto been totally obscure. Recently, however, direct hypothalamo-autonomic connections from the paraventricular nucleus to the central parasympathetic and sympathetic nuclei of the brainstem and cord of the rat, cat and monkey have been demonstrated by Saper *et al.* (262), using retrograde movement of horseradish peroxidase as a marker. Swanson (290) has also shown the presence of neurophysin in these connections in the rat and ox. These studies have brought the dorsal nucleus of the vagus, the Edinger-Westphal nucleus, the nucleus of the solitary tract, and the intermediolateral column of sympathetic neurones in the thoracic cord into direct relationship with the hypothalamus. Furthermore the carrier protein of the HNH system (neurophysin or a very similar protein) is involved. Thus we may postulate some involvement of this neurophysin system in scrapie, where we know that the main neurophysin system, emanating from the same paraventricular nuclei, is affected (see Chapter 8).

Other changes in the nervous system

Apart from the changes in the HNH and cerebellar systems, other (mostly inconstant) lesions have been observed, some systematised, in specific locations, others generally distributed through the brain (for details, see Beck *et al.* (23)). Fresh lesions were found in a variety of hypothalamic nuclei, amygdaloid nuclei, medial habenular nuclei, and elsewhere. Diffuse changes included thickening of meninges and of the walls of small blood vessels; ventricular dilatation; subpial and subependymal gliosis; diffuse satellitosis, and occurrence of Alzheimer type 2 glial cells. Inflammatory changes were inconspicuous (absent in 30 cases), but included occasional vascular cuffing with small round cells (in two of the clinically severest cases), while in another two there were multiple microglial nodules. In one case there were numerous PAS-positive plaques (see Beck *et al.* (23), Fig. 7) in an amygdaloid nucleus. Somewhat similar 'plaques' were subsequently observed by Bruce and Fraser (54) in mice with experimental scrapie. Their significance is obscure.

Certain negative findings deserve emphasis, in view of the generally acknowledged similarities between scrapie in sheep and two human diseases, kuru and Creutzfeldt-Jakob disease. The thalamus and corpus striatum, frequently affected in Creutzfeldt-Jakob disease, appear intact in natural scrapie; likewise the main motor and sensory pathways. Status spongiosus, associated with neuronal loss and intense gliosis, is characteristic of Creutzfeldt-Jakob disease and of kuru— especially of experimental kuru in primates. It is also seen in experimental scrapie in mice, but is practically never seen in natural scrapie (21). This finding agrees with that of Zlotnik (336, p. 125) in animals of the Cheviot breed.

The role of astrocytes

In the earliest preclinical stage, when there were degenerating mossy endings in the cerebellum without any other change, no astrocytic reaction at all was seen. This has been confirmed in known 'black' animals in the preclinical phase. Astrocytic proliferation occurs as a secondary non-specific reaction with development of the neuronal decay, and is associated with fibrous gliosis. Moderate astrocytic proliferation cannot be used as an indicator of scrapie, since it occurs in non-scrapie animals from the age of $2\frac{3}{4}$ years at least (19).

Endocrine glands

The pituitary gland. In scrapie animals the posterior lobe, when stained with the Gomori chrome-alum-haematoxylin (CAH) method, contains only scanty remnants of the normally very abundant CAH-positive neurosecretory material, while the density of its cells is considerably increased. The anterior lobe shows no abnormality. In the

median eminence and pituitary stalk, however, there is an increase of
CAH-positive material, in Herring bodies and in beaded fibres, far
in excess of the normals. The material is also present in the pars tuberalis, a phenomenon never normally seen in the sheep. Thus the changes
are confined to the neurohypophysis, with evidence of increased neurosecretory material in the proximal part and marked reduction in the
distal portion (see Chapter 8).

The thyroid glands. These are small and translucent and show a normal
flat basal epithelium without any obvious changes in the colloid or
follicles.

The adrenal glands. In longstanding cases these are much enlarged, owing
to increased mass and width of the cortex, which is a pale grey-white
instead of a light pink-brown. The increase is due to a great distention
of the zona fasciculata, the cells of which are much enlarged, with a
foamy cytoplasm, and tightly packed together.

The ovaries and testes. These have appeared to be normal.

The muscular system

The skeletal musculature of scrapie-prone sheep is frequently especially
well-developed in adolescence and early adult life, both in its mass and
its tonic state, as revealed by palpation of the back muscles.

In an early series of 45 scrapie-affected sheep (43) myopathic changes
were observed at autopsy in many sites, some of which in the hindlimbs
were very striking macroscopically. More frequently the affected muscle
showed foci of pallor or a general translucency resembling fish muscle,
with flabbiness and a marked atrophy, by weight and mass, of the whole
muscle (Fig. 7.9).

The distribution of the most striking muscle lesions varied very much.
Some muscles, such as the iliopsoas, the facial and auricular muscles,
the external eye muscles, and the muscles of the larynx, pharynx, and
tongue, were often extremely atrophic. These changes correlated well
with observed clinical disturbances; thus sheep which had great
difficulty in raising their heads had severe lesions of some of the dorsal
neck muscles; those with faulty eye movements severe lesions of the
extrinsic ocular muscles; some with gross ataxia of the hind limbs had
severe lesions in the muscles of the thigh and the pelvic girdle, where
the semitendinosus and semimembranosus muscles were often involved;
and so forth.

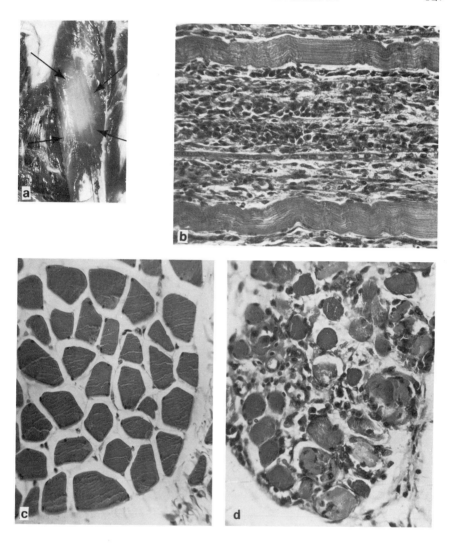

Fig. 7.9. (a) *Sharply demarcated focus of intense pallor (arrows) due to myopathy in* teres major *muscle of sheep with scrapie.* (b) *Longitudinal section from muscle shown in* (a) *showing two muscle fibres still retaining their cross-striations, with atrophic and disrupted muscle fibres between them, and an abundance of inflammatory and phagocytic cells. Iron haematoxylin and van Gieson, magnification* × 170. (c) *Transverse section of skeletal muscle of normal sheep, for comparison with* (d). (d) *Transverse section through affected skeletal muscle. Muscle fibres show great variation in size, and are in various stages of dissolution. Haematoxylin and eosin, magnification* × 250. (*Figs 7.9a–d are taken from Bosanquet* et al. (1956) *by permission of* The Lancet.)

Microscopically, the lesions were those of a severe myopathy, having features both of polymyositis and of muscular dystrophy (Fig. 7.9). For details, see Bosanquet *et al.* (43). There were no findings suggestive of denervation. At the time, the authors came to the conclusion that myopathy constituted the essential lesion in natural scrapie. Doubt was cast on this view by Hulland (157), who examined muscles from 16 sheep with natural scrapie, 14 with the experimentally induced disease, and 12 clinically normal animals, coming from Cheviot, Swaledale and Suffolk breeds. Minor muscle changes were found in both scrapie and control animals; in only one—a case of natural scrapie—was a severe myopathy found, resembling the changes described by Bosanquet *et al.* Myopathic changes were seen in representative muscles in 24 of 25 cases examined by Beck *et al.* (23), but with recognition of the significance of the changes in the central nervous system as a specific neuronal system degeneration, interest in the myopathic lesions has diminished. Macroscopic lesions have been observed not infrequently in brief post-mortem examinations, not specifically designed to reveal defects of skeletal muscle. In the light of the widespread hypothalamic disturbance, it seems more probable that the myopathic changes should be considered as part of the disease, but secondary to primary metabolic disturbances determined elsewhere.

III. Clinical Signs and Pathological Lesions

The five main clinical symptom-complexes of (1) metabolic wasting, (2) ataxia, (3) rubbing, (4) behavioural disorder and (5) autonomic disorder were described in Chapter 4, and their main pathological correlations noted above. The neuropathological lesions strongly suggest that scrapie is primarily due to degeneration of neurones within specialised neuro-anatomical systems, i.e. it may be considered an example of a 'system degeneration' affecting at least three anatomico-physiological systems, the cerebellar, the hypothalamo-neurohypo-physial and the central autonomic. The term 'system degeneration' is used, in human neuropathology, to denote a group of disorders in which there is a loss of neurones in specific parts of the nervous system, without inflammatory response, and with no obvious cause. Greenfield (134) and Spatz (281), in their studies of the spino-cerebellar degenerations of man, considered that degeneration proceeds from the peripheral end of the axon towards the cell body. The subject of 'dying-back' has been reviewed by Cavanagh (71, 72). A similar process certainly occurs (i.e. there is a retrograde 'dying-back' from the synaptic terminal) in the hereditary progressive retinal atrophy of the dog (223). The finding, in natural scrapie, of degenerating mossy fibres of the cerebellum as the sole pathological change in preclinical cases is consistent with such a concept. Thus, in scrapie, the initial lesion may be considered to be at the level of the presynaptic membrane and terminal axon of a neurone in the brainstem which then suffers a retrograde degeneration of its axon and cell body, with 'secondary' vacuolation of its perikaryon. Decay of the primary neurone may lead to further retrograde trans-synaptic degeneration, with initiation of a similar 'dying-back' involving a second-order neurone in functional contact with the primary neurone, and in turn developing a 'secondary' vacuolation of its perikaryon. There may also be some anterograde degeneration from the cerebellar glomeruli and the Purkinje cells to the dentate nucleus. The observed changes in the cerebellar system fit such an interpretation closely. In the HNH system the locus of the initial lesion may lie in the axon terminals of the distal neurohypophysis; though definite lesions are difficult to demonstrate, other evidence points to this occurring—for example, typical 'secondary' vacuolation in the neuronal perikarya of the supraoptic and paraventricular nuclei. At the same time, or earlier, there is increased secretion in the proximal neuro-hypophysis.

Sporadic vacuolation of neurones of the lateral and ventromedial hypothalamic nuclei and amygdaloid nucleus may be attributed to retrograde degeneration (see Swanson (290); Fig. 5), and it is possible that vacuolation of the dorsal nucleus of the vagus may depend on retrograde degeneration from the paraventricular nucleus via Swanson's tract. Whether degeneration in the central autonomic system commences from a primary initiating fault or is a secondary retrograde phenonomen from the hypothalamus remains unknown.

To summarise: natural scrapie appears to be a primary neural system degeneration, of 'dying-back' type, with three crucial sites of neuronal degeneration. Its characteristics are (1) restriction, especially in the earlier stages of the disease, of the degeneration to structures belonging to the cerebellar system (medial and lateral vestibular nuclei, external (accessory) cuneate nuclei, red nuclei and Clarke's columns), to the hypothalamo-neurohypophysial system, and possibly to parts of the central autonomic system; (2) bilateral and generally symmetrical distribution of the degeneration; and (3) sequential loss of neurones from these sites over probably at least 1/10 of a life-time, occurring without obvious morphological or functional disturbance to remaining neurones, and without any indication, at least by microscopical inspection, of why particular neurones are selected for dissolution.

In addition to the specific neurological defects there are inconstant changes in the muscular system. Some of the clinical signs, especially the abnormal eye movements, 'floppiness' of the ears, changes of voice, and in some degree the hindlimb ataxia, may be related to muscular weakness, consequent to a myopathy.

No explanation has ever been found to account for the fact that the animals rub themselves so severely. Examination of the fine nerves and nerve endings in the skin (Oppenheimer, 1957, personal communication) disclosed no abnormalities, while all attempts to find lesions in the peripheral nerves by histological examination of the spinal nerve roots and dorsal root ganglia have proved fruitless.

It is possible that in some sites the rubbing may be related to discomfort arising in muscles; but it is difficult to account thus for the compulsive nibbling of the haired extremities of the limbs distal to the main muscle mass, and of the poll of the head and over the nose. It is worth noting that in man muscle pain may be associated with irritability of the skin in cases of polymyositis, while skin irritation may also occur in various metabolic disorders, including diabetes insipidus.

Thus we have to recognise that the single pre-eminent clinical feature, which determines the diagnosis of scrapie for most observers, namely the rubbing, has as yet no physiological or pathological explanation.

Chapter 8

Ultrastructure and Histochemistry

The data presented in Chapter 7 strongly suggest that a primary neuronal degeneration in the cerebellar and hypothalamo-neuro-hypophysial systems, and possibly in the central autonomic system, is responsible for natural scrapie disease; and they direct attention to the *boutons terminaux* (the synaptic junctions between neurones) of these systems, as the possible location of the initial functional fault. Some insights into the possible ultrastructural characteristics of this initial malfunction have been provided by electron microscopic investigations (37, 38, 39, 40) and by immunohistochemical studies for the presence of neurophysin (184, 185, 230, 231, 232) on a series of affected and unaffected animals of the Suffolk breed. The evidence from these studies forms the basis of this chapter.

The earliest lesion revealed by light microscopy is the degeneration of mossy terminals within the granular layer of the cerebellar cortex (23, 19). Our attention was therefore directed to this location, where substantial numbers of affected *boutons terminaux* are commonly seen in the flocculonodular lobe of the cerebellum.

I. Electron Microscopy

The material came from five cases of natural scrapie, with three normal controls, all of the Suffolk breed. Full surgical and technical details are given by Bignami and Parry (39). Blocks were taken from frontal cortex, Ammon's horn, substantia innominata (nucleus basalis), lateral mamillary nucleus, supraoptic nucleus, ventral pons, inferior olives, lateral cuneate nucleus, cerebellar hemispheres, nodulus and vermis. Adjacent blocks were taken for light microscopy.

Membrane-bound vacuoles were found within neuronal cell bodies, dendrites, and axonal terminals in the medulla, pons, granule layer of the cerebellum, supraoptic nucleus, lateral mamillary nucleus, and the pyramidal layer of the hippocampus. These were present in all five cases of scrapie. No vacuoles were found in semi-thin and ultra-thin sections in the three normal sheep. The size of the vacuoles varied considerably, the smaller ones having a diameter of 1–2 μm. Many of the vacuoles contained filamentous structures and membrane-bound papilliform processes and vesicles, 100–200 nm in diameter, which appeared to be formed by inward budding of the membrane lining the vesicle (Figs 8.1, 2 and 3). In addition, membrane-bound packets of round virus-like particles, 35–50 nm in diameter, were found occasionally within cytoplasmic papilliform projections into neuronal vacuoles in the supraoptic nucleus of one sheep, and in the cerebellar cortex in two others. For further details, with illustrations, see Bignami and Parry (39).

A second alteration took the form of axonal swellings, containing accumulations of granular vesicles, 50–150 nm in diameter, haphazardly mixed with dense bodies, neurofilaments and neurotubules, within small nerve processes in the supraoptic nucleus, cerebellar cortex, pons and medulla of scrapie-affected sheep (for details and illustrations, see Bignami and Parry (40)). These swellings closely resembled the so-called 'Herring bodies' in electron micrographs of the human infundibulum (32), the pituitary stalk of the dog (329) and the posterior lobe of the bovine pituitary (96). Herring bodies are generally regarded as circumscribed swellings of nerve fibres filled with neurosecretory material (NSM), and are normally confined to the hypothalamo-neurohypophysial (HNH) system (102). They were classified by Dellmann and Rodriguez into three types, type I consisting of accumulations of granules of NSM, type II representing a degenerative stage of type I, and type III a subsequent regenerative stage. In our material from scrapie-affected sheep, we observed bodies resembling types II and III in the supraoptic nuclei, brainstem and cerebellum. They were not present in normal young and middle-aged sheep, but were seen, in corresponding situations, in the brains of senile animals (37). In summary, whereas intracytoplasmic membrane-bound vacuoles containing cytoplasmic processes appear to be specific for scrapie, the Herring bodies and similar changes in small nerve processes are not specific for scrapie, but can be found in senility, and in a variety of

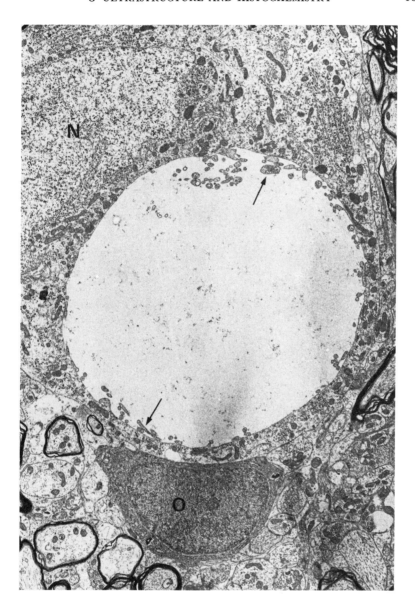

Fig. 8. 1. *Vacuolated neurone in the paracuneate nucleus of the medulla oblongata. Cytoplasmic processes extend like papillae inside the vacuole (arrows). Abbreviations: N, nucleus of neurone; O, oligodendroglial satellite. Magnification × 2800. (Figures 8.1, 2 and 3 are taken from Bignami and Parry (1972a) by courtesy of Oxford University Press.)*

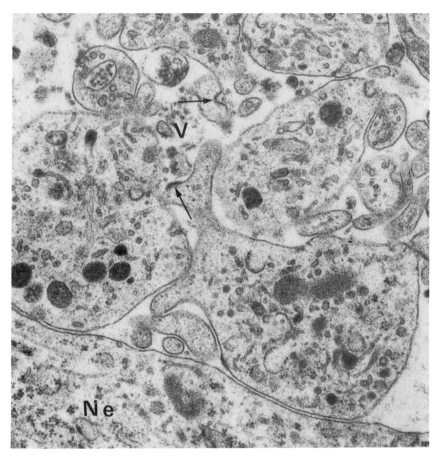

Fig. 8.2. *Cytoplasmic processes containing dense bodies, smooth tubules and vesicles within a vacuolated neurone in the supraoptic nucleus of the hypothalamus. Abbreviations: Ne, neurone V, vacuole. Arrows show fuzzy thickenings and bristle-coated invaginations of the limiting membrane. Magnification × 26,400.*

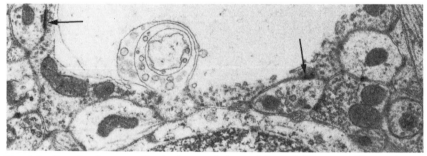

Fig. 8.3. *Vacuolated mossy ending in the granule layer of the cerebellar cortex. The membranous structure within the vacuole probably represents the detached vacuolar membrane. Arrows show synaptic membrane specialisations. Magnification × 16,000.*

toxic and degenerative conditions. Their significance is discussed by Bignami and Parry (40).

II. Immunohistochemical Studies and the Demonstration of Neurophysin

A more precise characterisation of the neurosecretion-like material found in the degenerating *boutons* was desirable for two reasons; (i) the discrepancies observed between the staining reactions of this material and the true neurosecretory material of the posterior lobe of the pituitary, with and without oxidation and after trypsin digestion (37), and (ii) uncertainty regarding the nature of the dense-cored vesicles revealed in degenerating neuronal terminals by electron microscopy. The fortunate availability of chemically pure preparations of neuro-physins prepared from pig pituitaries (154) and of highly selective antisera produced in rabbits against a specific neurophysin II (310), the amino-acid sequence of which was known, allowed the precise demonstration of neurophysin in sections of the sheep tissues, using the porcine neurophysin II antiserum, which proved to be species cross-reactive with the neurophysins of the sheep (see Livett, 183). The question was whether or not the neurosecretion-like material contained neurophysin, the protein-carrier of the pituitary hormones, vasopressin and oxytocin, and if so whether the neurophysin was restricted in its distribution to the specific neurones, and the specific *boutons*, where degeneration was occurring. For this, the method had to be capable of identifying individual neurones and individual *boutons*, a requirement not practicable with the conventional dyestuff histology, nor by bio-chemical analysis of blocks of tissue, but just possible with immuno-histochemical techniques (186), if sufficiently pure antigen were available to make a specific antiserum. It is convenient to use the following abbreviations: HNHS, hypothalamo-neurohypophysial neurophysin-containing neurosecretory system; SON, supraoptic nucleus of hypothalamus; PVN, paraventricular nucleus; NSM, neurosecretory material (as demonstrated by the chrome alum-haematoxylin-phloxine staining technique). The material came from 14 pure-bred pedigree sheep from three breeds; four with advanced scrapie, four moderately affected, one senile animal, and five normal sheep. Clinical, surgical and technical details are given by Parry and Livett (231).

In the course of this investigation it was observed (230) that, in addition to the well-known pathway leading from the large cells of the SON and PVN via the pituitary stalk to the distal neurohypophysis, and the pathway from the basal medial hypothalamus to the external infundibular zone of the median eminence and pituitary stalk (291, 140, 250, 247) there was a third pathway arising in the SON and PVN, leading to the proximal neurohypophysis, terminating in the external infundibular zone (see Fig. 8.4). The existence of this pathway has since been confirmed in a variety of mammalian species (for references see Parry and Livett, 231).

Neurophysin, as revealed by immunofluorescence, was confined, in the normal sheep, to the neuronal cell bodies of the SON and PVN of the hypothalamus, and to their axons. Fluorescent granules, showing a variety of patterns of aggregation, could be seen throughout the internal and external infundibular zones, in the pars tuberalis, and in all three portions of the infundibular process (posterior lobe of pituitary). It was present in the pars intermedia, and could occasionally be traced into the adjoining area of the adenohypophysis (anterior lobe). No specific immunofluorescence was found in the rest of the hypothalamus, or in other parts of the brain, or in the pineal body.

In the four animals with advanced scrapie, the distribution of neurophysin was markedly altered. In the SON and PVN the number of cell bodies showing specific immunofluorescence was reduced, probably to a third or a quarter of the normal. There was also a great reduction in the amount of neurophysin throughout the pathway leading to the *distal* neurohypophysis including the *internal* infundibular zone; in contrast, there was a two- to threefold increase in the *proximal* neurohypophysis, seen throughout the *external* infundibular zone. No immunofluorescence was seen in the pars intermedia and pars distalis. In the four less severely affected animals, loss of immunofluorescence in the distal neurohypophysis was readily detectable, but less severe.

Tests for the presence of the hormone vasopressin were made by assays of pressor activity at four levels of the HNHS in one normal animal and in one with advanced scrapie. Pressor activity was present at all four levels in the normal, and reduced by about two-thirds in the scrapie-affected animal. The difference was most marked in the distal neurohypophysis.

In the scrapie-affected animals, but not in the normal sheep, there was specific immunofluorescence in the cerebellar cortex. This was most

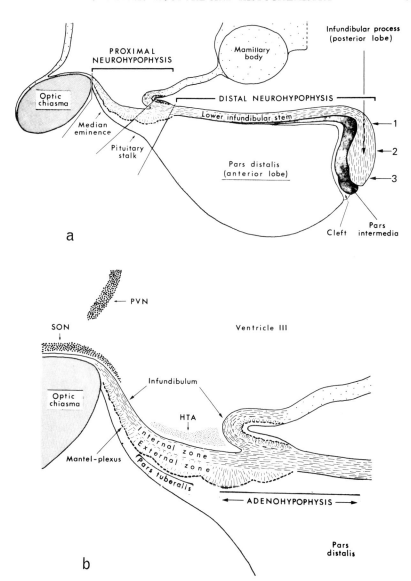

Fig. 8.4. *Diagram of the anatomical arrangements of the ventral midbrain and the pituitary gland in the sheep. (a) The main components as seen in the mid-sagittal plane with the three parts of the infundibular process: 1, the proximal cone; 2, the mid-lobe; and 3, the distal crescent. (b) The anterior portion of (a), to show in greater detail the structure of the proximal neurohypophysis. Although the diagram represents a mid-line section through the cavity of the third ventricle, the approximate sites within the hypothalamus of the supraoptic nucleus (SON), the paraventricular nucleus (PVN) and the hypophysiotropic area (HTA) are indicated. (Figure taken from Parry and Livett (1976) by courtesy of the editor of* Neuroscience.)

marked in the flocculonodular lobe (232). Some was also seen in the oculomotor nucleus and its outflow in one scrapie animal. No systematic search for neurophysin was carried out in the brainstem or spinal cord. The distribution of neurophysin within the cerebellar cortex was confined to the granular layer and the layer of Purkinje cells. The impression was gained that the material lay in or near the terminations of axons, within glomeruli or on the surfaces of Purkinje cells.

In parallel with the immunofluorescence studies, sections were stained by the Gomori chrome alum-haematoxylin-phloxine method for the demonstration of neurosecretory material (NSM). In general, there was close correspondence between the sites of immunofluorescence and those of Gomori-positive material. There were differences of detail, the reasons for which are discussed by Parry and Livett (231, 232). In the cerebellar cortex, there was a striking correspondence in the location of changes in the scrapie-affected animals, revealed by four separate lines of investigation.

(1) Degenerating *boutons terminaux* (23).
(2) NSM, as shown by the Gomori method (37, 231, 232).
(3) Herring bodies, as shown by electron microscopy (40).
(4) Neurophysin, as shown by immunofluorescence (231, 232).

A further change, revealed in precisely the same sites by non-specific autofluorescence and by electron microscopy, is an abnormal abundance of lysosomes and of lipofuscin granules (40, 232). The biological implications of these observations are discussed at length by Parry and Livett, (231, 232). These authors put forward a tentative hypothesis that in scrapie there is a disorder of protein metabolism in the central nervous system, in which sulphur-containing amino acids are particularly implicated.

Chapter 9

The Relationship of Scrapie to Other Diseases

The availability of sheep of known hereditary proneness to natural scrapie—the three genotypes of a single autosomal recessive gene—has allowed studies of the early preclinical pathology, which have yielded some insight into the possible underlying cellular metabolic disturbances. These observations may help to provide an explanation for the locations of neuronal loss in specific neuroanatomical 'systems' in natural scrapie, which are reminiscent of those found in the group of nervous diseases known as 'system degenerations' in man. A number of these neurological disorders show a heredo-familial pattern of occurrence, manifest at variable ages in adult life, and may exhibit a variety of clinical syndromes within a single family, such as that described by Schut (267). They may be considered as examples of Gowers's *abiotrophies* (133). Natural scrapie of sheep, on the basis of its natural history, pathology and epidemiology, may be placed in this general category without prejudice as to whether an infective agent or a genetic mechanism is primarily responsible for the abiotrophic decay. The term 'heredofamilial' is a useful designation for a disorder occurring in several members of a family or clan in several generations, with a pattern suggesting a Mendelian (i.e. nuclear) form of segregation compatible with heredity. If the word 'late' be inserted before 'heredo-familial', then the term covers the crucial biological aspect of these abiotrophic disorders of middle age—i.e. that they occur after normal reproduction has been achieved. In these, unless meticulous family and clan histories can be assembled, essential quantitative information may be lacking.

I. Late Heredo-familial Abiotrophies

The Basic Defect and the Temporal Delay Before Functional Deficits Develop

If we suppose that the basic defect may be genetic, as it is in a long series of congenital disorders of man and animals due to inborn errors of metabolism (126), then we must seek an explanation for the temporal delay or *latent period* before evidence of the presence of the abnormal gene appears, in the form of clinical illness. In most of the degenerative disorders there is no indicative 'marker' which may be detected by immunological, biochemical or physical tests, whereas such 'markers' occur in most of the congenital diseases involving a metabolic 'error', giving warning of impending disability prior to the onset of clinical signs.

From such evidence as is available, the basic defect may be taken to reflect a loss of the normal ability to maintain the functional integrity of a part, sometimes in a very limited and circumscribed area, *the site of primary lesion*, which undergoes physiological malfunction, anatomical decay or premature senescence, due to some metabolic inadequacy. An elegant example is provided by the Mendelian autosomal recessive form of generalised Retinal Atrophy of Red Irish Setter Dogs (223), where the physiological malfunction occurs within 48 hours of the physiological maturity of the retinal nerve cell, as judged by the development of an electroretinogram at 85 ± 2 days post-conception (gestation 56–65 days), which then undergoes changes and within a few days cannot be detected (233). This loss of the electroretinogram, which precedes by some months any clinically detectable visual defect, coincides with the first morphological changes detectable by light microscopy in the outer limb segments of the rod cells, all of which decay sequentially and 'fall out' slowly over the next 6–12 months (equivalent to one-tenth of the life-span) with progressive deterioration of night vision. The rate of decay may be easily measured quantitatively in histological preparations of the retina by counting the number of bands of nuclei in the thickness of the outer nuclear layer of the tapetal fundus, since the 'dying-back' and the 'fall-out' of the rod cells is symmetrical throughout the retina. The primary initiating lesion is in the rods, without morphological changes in the pigment epithelium or blood supply (223).

Thus the degeneration commences in the most peripheral or distal portion of the rod, the outer limb segment, and proceeds towards the cell body; no clue as to a possible metabolic anomaly, such as occurs in ornithine retinopathy (294, 295), is known. Similar 'dying-back' occurs in cerebellar system atrophies (134, 281, 71, 72). The finding of degenerating mossy terminals in the cerebellum in three sheep considered to be cases of preclinical natural scrapie (23), and confirmed by subsequent studies (37), supports the view of natural scrapie as a further example of a system degeneration or abiotrophy, in which the primary lesion arises in the most distal portion of the neuron and then proceeds centripetally to the cell body.

From the cell loss in the primary initiating lesion *secondary anatomical effects* ensue due to neuronal and transneuronal degeneration, mainly retrograde, as well as the disturbance of the physiological functioning of complex neural pathways and of the glial supporting tissues. The consequent functional deficits may be compensated for some time, certainly for months and sometimes longer, before the deficits prove so large that clinical manifestations ensue. Similarly in generalised progressive retinal atrophy (223), the neurones of the inner nuclear layer and of the ganglion cell layer gradually disappear over several months, leaving a few neurones with an intact cone cell layer. At this stage night vision is absent, but day vision remains for a few more months, after which vision is totally lost, and all the retinal neurones have disappeared and been replaced by glial tissue. In atrophies of the brain and spinal cord, the secondary effects, especially those due to transneuronal retrograde degeneration, are more widespread, as the primary initiating lesion frequently affects multiple loci, and glial and microvascular reactions are stimulated and involve adjacent structures.

One characteristic, but as yet unexplained, aspect of these abiotrophic disorders is the slow progressive and relentless loss of neurones in the primary lesion site, which undergo decay and disappear without any obvious residual evidence of their previous position. Thus assessments of neuronal loss in the early stages of a disorder are often unreliable without neuronal cell counts and may be unrecognised, unless gliosis be present. The initiation of the abnormal degeneration of some neurones is apparently proceeding continuously, while many other contiguous neurones remain morphologically unaffected and apparently function normally, or at least sufficiently well to afford no clinical evidence of functional deficits. Once initiated, the degeneration probably proceeds invariably to the total dissolution of the neurone without abatement or repair.

An understanding of the factors controlling the *rate of neuronal decay* may indeed be the central problem, for the disease may reflect either the acceleration of a normal process, or an abnormal one. Why an abnormal rate of decay should occur is in most instances obscure. Suggestions include a slow quantitative inadequacy of some essential metabolic process; a quantitative defect in protein synthesis leading to the formation of a slightly abnormal 'fraudulent' protein (241); a somatic mutation in supporting glial and vascular cells (59); or a premature switch-over of some metabolic controls which may occur normally at certain ages, such as occurs in the synthesis of haemoglobin at the time of birth (see Bunn *et al.*, 57). Indeed abiotrophy might properly be considered to be a premature onset of physiological senescence restricted to certain neurones and neuronal locations.

The widely-used term 'anticipation' for the occurrence of a disorder at a younger age in successive generations, for which no satisfactory explanation is generally forthcoming, may in fact be related to the phenomenon of control of rate of decay. Anticipation sometimes occurs in scrapie, but so, on occasion, does the reverse, i.e. delayed or late manifestation. Indeed the frequent invoking of the terms 'incomplete penetrance' and 'expressivity' to explain epidemiological data which do not quite fit predicted Mendelian segregation ratios may reflect a similar reduction of the rate of decay. In some instances, the pheno-menon may be ascribed to homo- or heterozygosity of the phenotype, but it seems possible that the general influence of other parts or even the whole of the remainder of the genome may be involved. The present difficulty is that without more precise information it is not possible to submit these hypotheses to precise experimental investigation.

The data on retinal atrophy in the dog provide the outline of a general principle of aetiology and pathogenesis, albeit in a rather special neural system. However, the recent demonstration of early 'preclinical' changes in the proximal hypothalamo-neurohypophysial system in sheep of natural scrapie genotype similar to those observed at an older age in affected animals (231) suggests that the same principle may apply to the neurological abiotrophies of the brain and spinal cord, whether the initiating cause be a 'slow virus' or a gene.

Hereditary Control of an Abiotrophic Defect

We may consider several primary ways in which the defect might develop under basic hereditary control: (1) *by the metabolic effect of a conventional chromosomal gene* showing simple Mendelian segregation, either as a recessive, as in generalised progressive retinal atrophy of dogs (223) and Friedreich's ataxia in man (244, 14) or as a dominant, as in some forms of human retinitis pigmentosa and Huntington's chorea (14), or on the X-chromosome as in Leber's optic atrophy in man (280, 114); (2) *by the secondary effects of polygenic interacting alleles controlling unrelated bodily systems* such as blood groups, cell antigens, immunoproteins etc., which of themselves exhibit no harmful effects, and indeed may, in the absence of the deleterious gene, confer some selective advantage; (3) *by non-nuclear cytoplasmic factors* (including viruses), which may be transmitted through the ovum (164, 210); (4) by a *chromosomal gene* controlling the individual's susceptibility to some environmental agent, which might be (a) a virus or (b) an *ingested toxic product* such as agenised wheat, causing nutritional disturbances in certain strains of mice (146) or (c) a specific *inorganic chemical substance* such as copper, of which there are abnormal accumulations in patients with Wilson's hepatolenticular degeneration (277, 279, 18); (5) by *somatic mutations* (59) in cells undergoing mitosis. In the central nervous system, this would apply only to glial and connective tissues, since it is generally considered that neurones are not able to replicate once their maturation is complete (246, 86).

Patterns of Clinical Manifestation

In some of the heredo-familial abiotrophies (e.g. Huntington's Chorea) the clinical picture remains fairly constant within an affected family. In others, the clinical manifestations are characteristically variable, with various constellations of signs and symptoms which often occur in patterns sufficiently distinct to form *syndromes* attracting their own specific names, and often in the past regarded as distinct disease entities. A classic example is the family described by Schut (266, 267), affected by several different forms of hereditary ataxia, with age of onset varying between early adult life and late middle age. Records are available on 343 individuals over six generations, with 45 affected members (269, 268, 197). Variation occurs not only in the age of onset (or rather of clinical manifestation), but in the duration of the disease and its rate of progression, and in the relative severity of the lesions in the selection of nervous structures undergoing the abiotrophic process, and in the resulting clinical disturbances. A very similar variability is found in scrapie.

For the study of abiotrophic disorders, animals afford a number of advantages over human patients. In the first place it is impractical, if not impossible, to keep an incapacitated animal alive for a long period by intensive nursing. In consequence, they do not develop the non-specific pathological changes associated with a long illness which in man become superimposed upon the specific changes due to the disease. Secondly, it is often the case, as has been shown in natural scrapie (19) and in experimental kuru and Creutzfeldt-Jakob disease (196), that the typical lesions can be demonstrated some time before the disease becomes clinically manifest. It is therefore difficult to estimate the true incidence of a familial disorder on clinical observations alone; yet little else can be achieved in human familial disease, except where pre-monitory markers, indicating a preclinical stage of the disease, are available. In animals, post-mortem examinations can be carried out whenever convenient, at any stage of the disease, and much may be learnt from pathological examination of apparently healthy relatives of affected individuals.

The Constitution of an Affected Population

The genetic definition of an affected or unaffected population may be extremely relevant to a proper understanding of the aetiology of these abiotrophies. In the present context the word 'constitution' is being used to designate the nature, origin and location of a population, its mating practices and constraints, and the degree of consanguinity and of in-breeding which may be present in the affected and unaffected families and clans. As much quantitative detail, for as many generations as possible, should be collected. If an unaffected sub-population of a main population is sharing the same location and environmental influences, similar data for the unaffected group should be collected wherever possible, and due note taken of environmental factors which may prove relevant. The occurrence of a group of cases of disease as a cluster in a limited location is commonly thought to indicate an environmental aetiological factor, but it may equally well indicate a locus of high frequency within an in-breeding or restricted breeding population of certain alleles, which influence the manifestation of the disease. Experienced breeders of livestock have been well aware of such influences for a very long time, and the best breeders in all parts of the world take suitable measures to limit any concentration of deleterious alleles. In human affairs similar biological constraints prevail, though they are not always admitted. The prevalence in recent years in Britain of the sociological dogma that environmental influences are of paramount importance in almost all aspects of human biology (303) has led to a general reluctance to admit the possibility of hereditary influences, and hence a reluctance to look for and collect evidence that might disclose their existence.

A not unimportant aspect of the study of scrapie in sheep is the contribution it may make to the general understanding of these abiotrophic disorders. The sheep is of approximately the same body weight as man, although its brain weight is only one-tenth to one-twelfth of man's, and it has a life span and inter-generation interval about one-tenth of the human. It has a placid temperament, which allows close clinical and neurological examination, which is supported by a vast fund of critical observation by the traditional shepherds. Its breeding may be accurately controlled, providing reliable pedigrees, while pathological investigations may be conducted whenever scientific enquiry requires them.

Study of the sheep and its heredo-familial disorders may yet make a contribution to genetic science of a kind which is becoming almost impossible in human populations, and in certain aspects never was possible. Sheep are available in relatively small interbreeding groups or breeds of 'fixed' phenotype, organised in subgroups or flocks, which are almost entirely isolated from contact with other sheep, not exposed to a multitude of virus infections, and situated in various physical environments, which provide control data on possible environmental factors. They are available for experimental studies. The main draw-backs are our lack of detailed knowledge of many anatomical and physiological features of the sheep (notably of its neuroanatomy) the difficulty of applying and interpreting standard metabolic tests, and the small number of trained scientific workers available and interested in the problems of epidemiology.

II. Scrapie and Disease in Man

Three human diseases—multiple sclerosis, kuru and Creutzfeldt-Jakob disease (CJD)—have been thought at various times to be linked with scrapie. One supposition is that the transmissible spongiform encephalopathic agent (TSEPA) of scrapie may be pathogenic for man, and that exposure to the agent from handling affected live sheep or their carcases, or from the ingestion of the agent in sheep meat and its products, may play a significant role in the aetiology of these diseases, the precise causes of which are at present undetermined.

To establish a convincing causal relationship between such a postulated exposure and a human disease which may not appear until 20–50 years later is a matter of great difficulty. I have for 31 years watched for any evidence of neurological illness in the community known to me and associated with sheep as shepherds, owners, slaughtermen and butchers. None has come to my attention. A visit in 1964 to Dr. Pickles of Aysgarth in Wharfedale, North Yorkshire, who had made a lifetime's epidemiological study of contagious illness in this isolated dale of sheep farmers, elicited no recollection of any case of neurological illness of these kinds in his practice. Yet there is scrapie among the sheep of the Dale at the present time, and good evidence that it has been present for many years. There can be few places where an isolated and closely knit sheep-keeping community has been observed so carefully for over half a century by a perceptive medical practitioner.

The case for associating *multiple sclerosis* with scrapie does not bear critical examination, and is generally considered to be quite unfounded. The neuropathology of multiple sclerosis, with its 'plaques' of inflammatory demyelination, which remit and then later recur in the same place or elsewhere, is quite unlike the pathology of scrapie. Although a virus has often been proposed as a cause, the type of agent postulated is generally quite unlike the rather peculiar TSEPA of scrapie. The geographical distribution of multiple sclerosis around the world (176) does not coincide with areas where scrapie is known to occur or to have occurred in living memory in the local sheep populations; for instance no scrapie is recognised in East Finland, Australia or New Zealand, where multiple sclerosis is fairly common. Apart from a single report, now rejected as resulting from a laboratory contamination, repeated attempts to transmit multiple sclerosis to experimental animals have failed.

The possible relationship between scrapie and the *transmissible spongiform encephalopathies* of man (kuru of New Guinea and CJD), is based on more substantial evidence. These disorders fall into a very distinct group on the basis of two characteristic criteria; firstly, their similar neuropathology, and secondly, the presence of a transmissible encephalopathic factor, which may be demonstrated by animal inoculation, after which a subacute spongiform encephalopathy, showing a very similar pathology, irrespective of the factor's origin, ensues.

The neuropathological resemblances between kuru and natural scrapie are particularly close, less so between CJD and scrapie. The encephalopathies which follow inoculation of recipient animals are likewise remarkably similar, although they differ in some respects from natural scrapie. The host-range for the agents from the different diseases is remarkably similar, with the scrapie agent causing disease in many New and Old World monkeys, as do the agents of kuru and CJD (see Marsh (193) and Gajdusek (124)).

There is therefore a strong *prima facie* case for considering whether scrapie, mink encephalopathy, CJD and kuru are not closely related conditions, caused by a single agent or a series of closely related agents or 'slow viruses'. We must assume that the TSEPAs in these diseases are the initiating cause of the pathological disorder, and not a secondary consequence of the disease process. As with scrapie, the evidence relevant to this point is scanty, and is not available at the present time for the originating species.

There are objections to the neat hypothesis of a common, closely related group of subacute spongiform encephalopathic viruses derived from the scrapie agent. The hypothesis is only plausible with some imaginative treatment of the evidence, as for example in Fig. 17 of Gajdusek (124), which sets out to show a common origin for all four conditions. The main objections are as follows.

(1) There is no evidence that the original outbreak of transmissible mink encephalopathy (144, 60) was due to the mink eating scrapie sheep carcases, scrapie being thought not to have been present in Wisconsin and adjoining states at that time.

(2) The view that natural scrapie is mainly due to eating affected placenta and embryonic membranes is certainly untrue, and even Pattison *et al.* (237), who proposed this route on the basis of one

laboratory experiment, would not expect more than 5% of all natural cases to be attributable to such a mode of dissemination.

(3) Any relation between subsequent human disease and butchery, shepherding and kitchen accidents with scrapie-affected sheep meat due to self-inoculation through the skin, or by ingestion of sheep meat and meat products, is extraordinarily difficult to establish. If it were of frequent occurrence, with a latent period of a few years, as with the recent instances of CJD which have followed surgical procedures (127, 108), one might expect instances to have aroused comment in areas where scrapie-affected sheep are slaughtered for human consumption, as has been common practice in England for many years.

(4) Any aetiological connection between the ingestion of scrapie-affected sheep meat with CJD will also be extremely difficult to establish, bearing in mind the many other environmental factors which may be involved during the many years before the disease commonly appears. The Israeli example of the Jews from Libya, who commonly eat the eyeballs and brains of sheep, and who have a prevalence rate for CJD of 30 per million as against 1 per million for Jews of European origin (145, 168) is striking, but a great deal more epidemiological evidence of the possible exposure, timing and amounts of contaminated products consumed will be needed to establish its relevance. These special eating habits of the Libyan Jews are shared with other ethnic groups in the region using similar sheep supplies. What is the attack-rate of CJD among them? The origin of the sheep-meat consumed in Libya is particularly complicated, since live sheep are imported for slaughter from many countries. It is also possible that Israelis from Libya may differ in their genetic constitution from their compatriots coming from Europe, as well as in the special environmental hazards to which their life-style exposes them. Discreet enquiries among connoiseurs in England of sheep's brains, for long considered a table delicacy until the economic changes of the past two decades, has revealed none with any signs of incipient dementia.

(5) The hypothetical connection of the kuru TSEPA to that of scrapie is even less plausible. Kuru was confined to the Foré and neighbouring tribesmen in the central mountains of New Guinea (124). This is an area of dense rain forest, in which ovine animals could not survive. It is difficult to conceive how the scrapie TSEPA or its ancestral form in sheep of several millennia ago reached this area of New Guinea, to which access has been notoriously difficult until the past two to three

decades of mechanised transport. How did the Foré people become associated with an agent known only in sheep of European origin? It is just conceivable that sheep meat may have reached parts of New Guinea from Australia since 1800, but scrapie has never been recognised in Australian or New Zealand sheep. The possibility of a consignment of scrapie-contaminated meat reaching the remote Foré country before 1940 in time to produce the kuru epidemic found by Gajdusek and Zigas in 1957 (125) seems unrealistically remote. Would it not be easier to suppose that some early European missionary visitor to the Foré region succumbed to the kuru-like form of CJD now recognised in Europe (271, 175), and, from desire not to lose his strength and influence, became added to the local food chain by cannibalistic rites?

[*Editor's postscript:* The existence of a group of diseases in the aetiology of which both genetic factors and potentially infectious agents may play a part raises questions which, though still unanswered, are at least beginning to be asked. During the past few years successful transmission experiments have been reported from cases of CJD in which the natural disease behaved as an autosomal dominant trait (255, 42, 195, 194) and from cases of familial Alzheimer's disease (305). Even more surprisingly, it has been suggested (309), that a virus may be involved in Huntington's chorea—the type disease for a primary neuronal degeneration with autosomal dominant inheritance. The issue is discussed by Oppenheimer (213), Traub *et al.* (305), Haltia *et al.* (141), Masters *et al.* (195, 194), and others. The answer to the problem of how a disease can be at the same time 'genetic' and 'infective' has not been settled; but enough has already been said to upset the simplistic view that because a disease is transmissible it cannot be inherited.]

Chapter 10

Conclusions and Recommendations

I. The Agricultural Importance of Scrapie

An annual scrapie attack rate of 10% or more is an economic disaster, now as in the 18th century, and can lead to rapid financial failure and discontinuance of the flock. Many pedigree flocks with a 1–3% annual loss can absorb such a wastage by prudent management. In the past, however, such a loss, inexplicable and uncertain as to its size and its future impact, has depressed and deterred shepherds and owners. In pure-bred flocks the importance of scrapie lies in loss of confidence, as well as in disruption of long-term breeding and improvement plans.

The impact of scrapie in *commercial ewe flocks* is probably the most serious aspect of the disease, whether these flocks are of a pure breed or cross-breds. The economic consequences of early loss and the cost of replacement rarely permit an annual ewe wastage rate from all causes exceeding 5%, except for an odd year or two. The maximum which may be allowed for scrapie lies in the 2–3% range. In the past decade many traditional lowland purchasers of half-bred ewes from the Pennine hills have had consignments showing an 8–10% annual loss rate from scrapie, which is quite unacceptable. The market demand has declined sharply, since it is nigh impossible, as it was in 1754 (167), to trace the breeders' source flocks and to obtain any satisfactory evidence of the scrapie-status of these flocks.

In commercial flocks producing meat animals for slaughter before 18 months of age, in which none will be retained for breeding, there are good grounds for using sires known or suspected to be carrying the scrapie allele. The progeny of 'black' Suffolk rams show a more rapid live weight gain in adolescence (284) and are generally more acceptable to the meat trade. The progeny of 'grey' rams show in general similar characteristics. As the generality of commercial flock ewes have a very low scrapie-allele frequency, these matings will produce the heterozygous 'grey' lambs which are economically advantageous as butchers' animals.

The breeding policies adopted in the small groups of *élite stud flocks* producing the most desired rams in each breed hold the key to scrapie in their breeds. The attention paid to scrapie depends on breeders' perceptions of its likely importance and economic consequences to themselves. Any breed in which scrapie has appeared puts the main flocks at risk, and the most careful scrutiny should be maintained. Two recent instances will suffice to illustrate this risk, which is particularly serious to breeds producing flock ewes for sale.

The Dorset Horn breed has been considered free of scrapie in this century. In 1958 a case of scrapie occurred in a Dorset Horn ewe of unknown origin in East Anglia. In 1965 a case was confirmed in a research institute flock the origins of which could not be ascertained (182). At that time the SDA included a Dorset Horn and Poll Dorset breeder-group, which was warned of the potential risk, but the group disbanded in 1969 under the blandishments offered by the Meat and Livestock Commission. Scrapie has since occurred in a former SDA member's flock in the early 1970's, and has led on to the disbandment of the County's Farm Institute flock, which was recognised as affected in 1977–78. There is little doubt that, with continued surveillance backed by reliable pedigree recording, this unfortunate outbreak could have been mitigated and contained before it reached the Farm Institute.

In the mid-1960's sheepmen of the Middle Pennines sought advice on scrapie, which was appearing in their area, and was suspected in the half-bred females sold into the lowlands, which are a very important element of the cash-income of their sheep enterprises. Scrapie was present in ewes of at least two hill breeds supplying sires of the half-bred commercial ewes. As the full pedigrees of the rams of the two hill breeds, even of the élite ram-breeding flocks, were very difficult to obtain, owing to the terrain in which the sheep were kept and mated, there seemed little chance of identifying flocks with low and high scrapie-allele frequency and applying a constructive programme to reduce the general breed frequency of the scrapie allele. The two breeds supplying the rams for cross-breeding were kept on low ground, and there seemed to be a chance of reducing the scrapie-allele frequency among them, and even carrying out a test-mating programme such as was already in progress in the Suffolk breeder-group. The practical difficulties of test-mating in hill breeds are greater than with lowland breeds, since one needs to identify and maintain observation of the female stock sold away from the area. In 1964 the sires of 12,000 lambs were identified from the lambs' ear-tags, and a further 7,000 in 1965; and a collaboration was arranged to ascertain the lowland farms to which they went. A number of these animals in Lincolnshire were observed and found to develop scrapie. In at least one instance the sire was identified and found to have scrapie, and the flockowner informed.

However, although sheepmen were advised of the likelihood of an epidemic of scrapie developing over the next decade, the financial stringencies following a catastrophic fall of prices in the autumn of 1965 rendered any scheme of recording and test-mating out of the question, and the limited recording of lambs had to be abandoned. The 'epidemic' has regrettably developed (58) among Masham sheep. Other cross-breds such as 'mules' are now also at serious risk.

In breeds producing sires for prime lamb carcases the consequences of scrapie are less obvious and in general less serious. They affect primarily the economic viability of an individual stud flock, through premature loss of breeding stock and through loss of reputation in the market place, reflected by demand and price. To commercial buyers heterozygous animals are at an advantage, as are the double recessives if they live to work for three seasons or more. The main problem is the maintenance of a low scrapie-allele frequency in the flock ewes compatible with acceptable phenotype standards, and to effect a balance such that scrapie in sold stock is sufficiently uncommon to avoid comment, while any early wastage in the home flock ewes is covered by enhanced receipts.

Exports

Many countries are now enquiring about freedom from scrapie, generally applying requirements which are not, in our experience, helpful or relevant. Certificates of freedom based on casual veterinary inspection are largely worthless. Affidavits of freedom by owners are of limited value, especially when purchases are made through dealers who can honestly plead ignorance of any unfavourable facts. The SDA has had a number of frustrating experiences. On several occasions during the past 15 years it has been approached by representatives of overseas buyers for animals with as low a scrapie risk as possible. Animals have been drawn out for inspection and accepted as suitable; in every instance, although agreement was reached at the time of inspection, no sale has followed; and it has subsequently been learnt that purchases have been made from scrapie-affected flocks with uncertain records, at prices 10–15% below the SDA price. This difference is not unreasonable for a special product, and a very small item on the total cost of importation.

The result is that many exportations from Britain since 1950 of sheep of breeds known to carry the scrapie trait have given rise to outbreaks of scrapie, some recorded and some almost certainly not. If the sheep are kept within registered pedigree flocks, little harm may ensue. However, if the plan is to use rams of scrapie-affected breeds to improve the indigenous unimproved ewe stocks, then much stricter criteria are called for. Such importations have been made to Colombia, and are projected for Kenya. One breed, which has succeeded in exporting scrapie to South Africa several times (315) and to Kenya once (83), has in recent years done so to Colombia and Brazil. A request from Kenya for a source of scrapie-free sheep of this breed for use on local native sheep has provided an opportunity of advising on the risks involved, since no sheep of this specification can be identified in Britain or in Europe at the present time. Such an importation could well lead to disastrous consequences in 10–20 years' time under the more difficult conditions of African husbandry.

II. The Control of Scrapie

Measures to limit the incidence of scrapie may be considered conveniently at three levels: first, that of the individual sheep-keeper; second, that of the breed or breeding group; and third, that of national agencies. In the United Kingdom and until very recently in most of Western Europe except Germany, there have been no national government programmes, and it is doubtful whether at the present time any useful purpose would be served by having one in Britain, where any form of notification is likely to lead to suppression of information, and to many disputes as to diagnosis, which could be settled only by sophisticated and expensive pathological services.

For the individual flock keeper there is little to add to the advice freely available in 1800; that is, he should be careful to buy stock from established flocks free of the scrapie trait and from flockmasters of repute. There is no need to dispose of affected animals until it is convenient to do so. If the flock is breeding its own flock replacements, greater care should be given to the selection of rams (cf. Fink, 118), and some form of parentage recording kept. Unless required for a test-mating programme, no F_1 progeny of affected parents should be retained; if one parent is affected, most of the F_2 progeny should be disposed of. The progeny of an unaffected parent which has had (or subsequently has) an affected offspring should be kept under close observation, and if evidence of scrapie appears in related individuals the owner must be prepared to cull the whole family. In the selection of breeding stock he should select those with parents over $4\frac{1}{2}$ years old, or with grandparents older than $4\frac{1}{2}$ years, unless there is a sound non-heritable cause for earlier death. If animals with younger parents are retained, as is often desirable, he should keep in touch with the parents and be prepared to cull if any sign of scrapie appears. The severity of the culling has to be balanced against disruption of the flock breeding programme; sometimes the risk of scrapie may be the lesser evil.

Flocks supplying female breeding stock

The individual commercial flock owner is in practice dependent on a supply of breeding stock free or virtually free of the scrapie trait, i.e. of low scrapie-allele frequency. This is the second level, of the breed or breeder-group, which is the nub of the whole problem. If the élite ram-breeding flocks of a pure breed are virtually free of the scrapie allele (and many breeds are), no scrapie, or very little, occurs in the secondary flocks in the breed or in their first crosses. In breeds in which scrapie is known to occur, the incidence of scrapie is usually capricious, and in general one can only rely on intelligence by whispered word of mouth, often in code—a most unsatisfactory and incomplete method of data collection, but it has to suffice. Breed society records are of little help, but one must consult parentage, production and health records, which may be available for any source flock, and build up full details of the life-time history of any rams, which may seem suitable for one's purpose, taking note of general productivity factors as well as of scrapie. In general one should purchase one's stock from flocks with detailed records made in tamper-proof bound shepherds' books, and set out in a form of flock register which may be easily consulted and checked, and from which entries cannot be removed without evidence. In breeds where a serious attempt is being made to produce scrapie allele-free stock, these should always be preferred, especially for females. The effective control of scrapie depends, in the long term, on the co-operation of the leading flocks in any breed, to establish the exact scrapie status of its most desirable rams, and to breed rams of these qualities free of the scrapie allele.

Where commercial half-bred ewes are being produced, it is only necessary to ensure that the ram does not carry the scrapie allele to produce unaffected flock ewes. At present, regrettably, we have evidence of scrapie in the Masham, Mule, Welsh Half-bred and some Scottish Half-breds (not in those with North Country Cheviot dams), but none in Kent or Romney Half-breds. If it proved possible to reduce the scrapie-allele frequencies of the Teeswater, Blue-faced Leicester and Border Leicester rams being used, scrapie would virtually disappear from these commercial ewes within a decade, as the Scottish Half-bred breeders have demonstrated in the past. An alternative method is to use rams free of the scrapie allele such as Suffolk 'proven whites', Dorset Downs, Southdowns, or even the less popular longwools such as the Cotswolds, Romneys, Lincolns and Leicesters, all of which are probably virtually scrapie-free, to produce flock replacements based on three breeds. Of these the Suffolk cross, especially on Scottish half-breds, the so-called Baumshire, is a well recognised and successful flock ewe.

Since the control of scrapie is, in our view, essentially a breeding problem, the present solution to scrapie lies in the hands of practical breeders. The means are now available, clearly defined and tested in practice over 20 years. The use that any breed or breeder-group sees fit to make of these will, of course, depend on the extent of the losses from scrapie and the economics of the breed. What has been possible in the Suffolk breed over the past 25 years would not have been possible in the smaller and less economically viable Southdown and Hampshire Down breeds. The size of the problem in the Middle

Pennines at the present time might well warrant a programme as sophisticated as the present Sheep Development Association's Suffolk scheme. Such a scheme may well require some support with test-mating, but it must be absolutely confidential, voluntary and free of bureaucratic control to ensure complete collaboration of flockowners and shepherds based on mutual and accepted shared responsibility of everyone involved; suppression and misinformation can easily wreck years of careful work.

While the place of governmental interest in Britain, with its long history of widely scattered sporadic distribution of the disease, has been restricted to benign encouragement of self-help by the industry, in countries thought to be free of scrapie the introduction of animals which develop the disease raises questions of an entirely different order. The disease is now generally treated by slaughter of affected and in-contact animals, with rigorous quarantine of surrounding areas of flock contacts. Such a policy is entirely appropriate for controlling either a contagious 'slow virus' or a deleterious gene, both of which should be effectively eliminated. This policy has been applied in Australia, New Zealand and Kenya on introduction of the disease, and recently in Germany, where *die Traberkrankheit* had virtually disappeared. The United States since 1952, and to a lesser extent Canada, have applied a similar policy under conditions less amenable to such controls, since the disease may well have been present, but unrecorded, before the present extensive slaughter and quarantine arrangements were imposed. These have been accompanied by such a substantial reduction in their national flock that the economic impact of the regulations has led to serious questioning of the programme's acceptability.

Sire Genotype Selection for Freedom from the Natural Scrapie Allele

In populations in which scrapie is an economic problem the only method which has been applied under conditions supplying accurate information over a period of years is that of the Sheep Development Association and British Sheep Society in Great Britain, based on the selection of rams free of scrapie, i.e. sire genotype selection for freedom from the scrapie allele (226, 228). After 20 years of testing Suffolk rams of stud quality for freedom from the scrapie allele, and their use in flocks of the breeder-group, the disease disappeared from three very severely affected flocks, and from the breeder group as a whole in 1977–1978, without selective culling of affected animals (228; Appendix A). The flock owners and shepherds are satisfied of the efficacy of this method in controlling scrapie, and wish to continue the programme in spite of the cumbersome nature of the test-mating procedure. Their main concern now is the development of families and strains of very low scrapie-allele frequency, which will be phenotypically as attractive as the most desired animals in the breed. It is of interest that the epidemic of 'goggles' (scrapie) in Wessex and East Anglia in the late 18th century disappeared under a similar sire genotype selection programme. On that occasion another scrapie-free breed, the Southdown, was used to provide the sires (95).

Sire Genotype Selection for Resistance to Inoculated Scrapie

Another investigation is in progress, based on the challenge of a basic nuclear ram-producing flock of Swaledales by inoculation with a scrapie brain-derived injection, to establish their response to the scrapie agent. Resistance to the inoculation is thought to be controlled by a single dominant gene (212, 150). Resistant animals and their progeny are being multiplied to provide a supply of rams for use in Swaledale flocks in the field.

It will be a decade or longer before one can expect any clear indication of the efficacy of this project. It faces substantial difficulties. First, the grounds for assuming that response to artificial inoculation is a sufficient indication of proneness to natural scrapie may be questioned. In other breeds, in our experience, sheep believed to be free of the natural scrapie allele are susceptible to inoculation with freshly harvested material from natural scrapie-affected sheep. Second, foundation animals for the nuclear ram-producing flock were donated from a large number of Swaledale flocks, and it is probable, in view of the high incidence of scrapie in certain sectors of the breed, that at least some animals will have carried the natural scrapie allele. In such a closed breeding population cases of natural scrapie are likely to occur, and make the evaluation of results difficult. Third, when a strain of inoculation-resistant sheep has been identified and multiplied, they must be acceptable, on breed phenotype characteristics, to the important Swaledale ram breeding flocks. Unless one is particularly fortunate, the development of strains of sufficiently desirable rams can prove difficult. Fourth, the evaluation of the efficacy of these inoculation-resistant rams in controlling natural scrapie on the hills will require very special attention and supervision. There are very few Swaledale flocks able to guarantee single ram mating groups. Owing to the nature of their holdings, individual animal recording is rare, and losses of sheep on the open hill difficult to monitor. One would hope that special support will be given to obtain the epidemiological data which are necessary to establish that any effect observed is due to the use of resistant rams and has not happened spontaneously. It is a very difficult undertaking.

III. Concluding Remarks

The state of our present knowledge of natural scrapie in Britain, outlined above, affords a provisional and tentative assessment. It should be regarded as a contemporary 'base-pad' of perceived knowledge, based on practical experience, from which more precise studies may be launched. We now know rather more what questions should be asked and what data collected. Such studies pose very real practical problems because of their long-term character and the difficulties of interpretation of apparently conflicting results, which have so frequently been 'smudged' in published reports. One awaits particularly a method of identifying the present, past, or even future contamination of an individual sheep with the scrapie agent (TSEPA) and firm evidence that the TSEPA is a primary cause of the natural disease. Is scrapie ever communicable under natural on-farm conditions, other than as a rare cannibalistic event? Just how good is the evidence from Edinburgh (100) for the vertical maternal transmission of the agent? And why do the results of the experimental studies from Mission, Texas (155) appear to contradict our own?

In the meantime, the recessive gene hypothesis offers the most precise quantitative account of the disease yet available, and has provided the theoretical basis for an effective means of control of manifestation of scrapie in one breeder group under British farming conditions.

Appendix A

[*Editor's note:* this article was submitted for publication in March 1978. The accompanying letter said:

> In view of the wide interest aroused in the general subject of the Transmissible ('Slow Virus') Dementias of Man and Animals by Dr. Gajdusek's Nobel Prize Oration last year, it seems important to place on record facts regarding scrapie in sheep which are generally overlooked in the scramble to establish a primary infectious aetiology for this group of disorders. The implications of these results are also important for the policies of Government Veterinary and Agricultural Departments around the world, who view scrapie, quite properly, as a potentially very serious matter.

The article was at first rejected on the recommendation of a referee, who considered that Parry's views and conclusions were erroneous. It was later accepted in a much abridged form, under a different title (228). It is reproduced here in its original form. The reference numbers have been altered to conform with the bibliography of the present volume.]

Natural Scrapie in Sheep:
Its Natural History and Disappearance Following Genetic Procedures

H. B. PARRY

Nuffield Institute for Medical Research
Headley Way, Headington, Oxford, U.K.

Natural scrapie is controlled primarily by an autosomal recessive gene. Identification and use of rams of double dominant genotype has provided progeny all unaffected, with the disappearance of scrapie in the breeding population. No natural communicability of, or aetiological role for, the artificially transmissible scrapie agent ('slow-virus') has been detected.

Scrapie disease of sheep is a slowly progressive fatal nervous disorder characterised by an age-related onset of clinical illness, usually in middle age between 2 and 5 years (225, 226) due to a non-inflammatory, bilaterally symmetrical, loss of neurones, in specific neuro-anatomical 'systems' (23), notably the hypothalamo-neurohypophysial and olivo-ponto-cerebellar. An artificially transmissible spongiform encephalopathic agent (TSEPA) or 'slow virus' is present when pathological lesions are established in the brain, but its role in the natural disease is unclear. The disease incidence is very variable and frequently shows a heredo-familial pattern suggesting an hereditary component.

Previous communications (225, 226) provided evidence that the clinical manifestation of natural field scrapie in sheep showed a pattern compatible with its determination by an autosomal recessive gene, irrespective of close contact with affected animals. The recessive gene hypothesis has allowed the construction of quantitative models of predicted cohort and flock attack-rates (106) which could be tested by direct observation, whenever precise knowledge of the sire genotypes was available. The identification of rams of double dominant (scrapie allele-free) genotype by a test-mating procedure has now allowed the testing of these predictions.

Methods and Material

Diagnostic criteria

In the absence of any specific laboratory test for the disease (121), for identifying the three presumptive genotypes, or practicable means of demonstrating the TSEPA, the diagnosis of affected and unaffected has been based on (1) clinical signs (226), (2) the family and clan health history (226), (3) autopsy after euthanasia (23), (4) histopathological assessment, using light microscopy on sections from nine levels of the brain and upper spinal cord stained by five or six methods and carried out 'blind' by one of three observers (23) and (5) occasional special studies using electron microscopy (37–40) and histo-immunochemistry (230–232). Of about 3000 affected animals, 2100 have been observed by the author and 1900 autopsied by him, with 1040 subjected to detailed histopathological study.

Classification of animals

Since about 85–90% of the observed cases of scrapie have occurred by $4\frac{1}{2}$ years old (226), this age has been taken as the minimum age for the main classification, determined as follows: (1) 'black', affected or unaffected with both parents affected, i.e. of presumptive homozygous recessive genotype; (2) 'grey', unaffected but with one parent affected or with affected progeny, i.e. of presumed heterozygous recessive genotype; and (3) 'white', unaffected with neither parent affected and with no affected progeny, i.e. of probable homozygous dominant genotype, although a proportion, the progeny of 'grey' × 'grey' matings, will be heterozygotes; (4) *'Proven whites'* are rams shown by test-mating to be free of the presumptive scrapie allele, i.e. of 'proven' double dominant genotype. Their progeny generations, sired successively without interruption by such 'proven' rams, are designated F_1W, F_3W etc, to indicate the halving of the scrapie-gene frequency in each generation. Assuming the original F_0 dams to be of the three genotypes in equal numbers with a gene frequency of 0·5, then the chance that an animal of the $F_{5-6}W$ generations carries the scrapie allele will be about 1 in 100, and of the $F_{8-9}W$ generations about 1 in 1000.

The population under observation

Stud flocks in Britain willing to keep meticulous identification, health and reproduction records have collaborated (272).* They retain at age $1\frac{1}{2}$ years about 35–40% of the female birth cohort but only 1–2% of the male, about half of each sex being available at ages beyond $4\frac{1}{2}$ years. The main data have been obtained in one breeder group of 15–20 flocks with 2500 ewes recorded continuously for 25 years, and in particular three flocks of 700 ewes in which all important progeny have been kept or sent to an experimental flock. Observations over shorter periods are available for 10 other breeds, thus providing data on some 40,000 individuals.

Test-mating for absence of the recessive scrapie allele

Rams have been mated with 'black' and 'grey' ewes to provide sufficient progeny for at least 10 offspring out of 'black' ewes or 22 out of 'grey' ewes, or a suitable proportion of each dam category, to be observed to age $4\frac{1}{2}$ years or later. Using a manifestation level of 76% based on recent, more stringent criteria (107), the probability that an allele-carrying ram will fail to produce in such a sire-progeny group an offspring affected by age $4\frac{1}{2}$ years is less than 1 in 100†. Such a ram with no affected progeny is assumed not to carry the scrapie allele and is termed a 'proven white'.

Accuracy of ascertainment

The principal error may lie in the small proportion (20–25%) of the zygotes produced actually observed to postnatal age $4\frac{1}{2}$ years. However, 'black' and 'grey' animals tend to be retained, which will increase above expectation the proportion of allele carriers in the observed birth cohorts (226); no

---

* See editorial note on p. 73. (Ed.).
† See note on p.98 (Ed.).

preferential intra-uterine, perinatal or postnatal mortality of the 'blacks' has been detected.

The diagnoses based on clinical, familial and pathological criteria have conformed closely with each other. In only two of the 1040 animals examined histopathologically did that assessment disagree with the clinical. Of 336 consecutive clinically affected animals 334 were fully positive histologically with two probably positive, while the control series of clinically normal animals from the same flocks and birth cohorts of 101 unaffected animals showed 93 negative, with six probably and two definitely positive, the eight being older than 5 years and probable allele carriers. Thus a positive clinical assessment is decisive, but is likely to underestimate potential cases in older animals.

The identification of natural scrapie allele-free rams

105 rams have been tested in eight flocks, of which 59 have had more than one affected progeny and have been discarded. Forty-six have had no affected progeny; of these 17 have proved free of the allele at a level of $p < 1:100$ or greater, four to $p < 1:20$-100, four are presently under test while for 17 the data are insufficient to provide an assessment and in four cases the test-mating was abandoned for other breeding reasons.

The 17 rams 'proven' free of the scrapie allele were born and reared in nine flocks, and have been at stud in three flocks, which have had scrapie present throughout the rams' lives. Ewes from 15 flocks, some affected with scrapie and some not, have been sent to these rams for mating, so that sub-populations of their progeny might be distributed throughout the breeder group.

The first ram 'proven' free of the scrapie allele had both parents presumptive heterozygotes, ten were sons of four 'proven' allele-free rams, while relevant information on the remaining six rams from six different flocks is lacking.

The Performance of the Progeny Generations of 'Proven' Allele-free Rams

The absence of scrapie in the F_1W generation

1223 progeny, 1167 female and 56 male, of the 17 'proven white' rams have been born and reared to adults in 15 flocks. 825 females and 12 males have been observed to $4\frac{1}{2}$ years old and older, many to 7 years, the oldest being $13\frac{1}{2}$ years. No case of scrapie has occurred in these sheep, although all 15 flocks have had cases of scrapie during some part of the observation period 1958–1977, 12 in their flock females, while in three flocks scrapie has been present continuously until the last 2–3 years.

Sixty-three progeny were out of dams affected at the time of their gestation or suckling, or manifesting scrapie within 2 years. Of the 63, 46 have been observed to $4\frac{1}{2}$ years old or older, with 25 to $7\frac{1}{2}$ years and some to 11 years of age; no clinical signs of scrapie have appeared, and histopathological examination of eight of the oldest animals, which had spent all their lives in contact with clinical scrapie, showed no evidence of scrapie or of the non-specific senile changes commonly found.

The re-appearance of scrapie in the F_2 back-cross generation

In the high-incidence flock cited in Table A.1 matings of two heterozygous F_1W rams with known allele carriers provided 21 progeny observed to $4\frac{1}{2}$ years or older. One F_1W ram out of an affected ewe was mated with (i) five F_1W half-sisters, daughters of the same 'proven' ram out of affected ewes; one of the six progeny became affected; (ii) three 'black' ewes subsequently affected, when two of the four progeny became affected; and (iii) six unrelated 'grey' ewes when none of the six progeny became affected.

The second F_1W ram with an unaffected dam produced one affected of three progeny out of 'black' ewes, and one affected of 14 progeny out of 'grey' ewes.

Observations of the F_2W–F_6W generations

308 females and 18 males of F_2W–F_6W progeny born in two flocks are under observation, with a small number of adolescent animals of the F_7W generation. No sign of clinical scrapie has been detected to date, although the final classification of the F_5W and F_6W generations is incomplete.

One allele carrier has been detected in the F_5W generation. A ram born in 1971 was test-mated in 1971–1972; one daughter out of a 'grey' dam has developed scrapie in 1976. The ram's great-great grandmother, born in 1956, was affected in 1959; from her daughter, born in 1958, the allele has apparently been passed 'silently' without clinical manifestation for 18 years through four generations of females, 10 offspring of which have been observed to $4\frac{1}{2}$ years or older without manifesting scrapie.

Distribution of Scrapie Among the Progeny of the Three Genotypes

With the identification of proven allele-free rams the schedule of reciprocal matings has been completed (Table A.1). The classifications of the presumptive genotypes of the sires and dams in Tables A.1 and A.2 have been based on parentage and other data, which are independent of the performance of the offspring assembled in the Tables. The results in section A were obtained in one flock with a continuous high scrapie incidence, while those in section B were obtained in flocks free of clinical scrapie among home-bred females. Other data are consistent with these results.

Table A1. *Occurrence of scrapie by age 4½ years among the progeny of the nine reciprocal matings of the three presumptive scrapie genotypes*

Number of sires	Parents' genotypes M F	Total of affected plus unaffected to 4½ years	No. of progeny Affected with scrapie Observed	Expected†
Section A				
7	ss × ss	56	48	48
	ss × Ss	124	57	53
11	Ss × ss	46	23	20
	Ss × Ss	106	18	22
7	SS × ss	39	0	0
	SS × Ss	7	0	0
	SS × SS*	159	0	0
Section B				
8	ss × SS*	156	0	0
2	Ss × SS*	23	0	0

* Presumed very low s allele frequency on family, flock and breeder group records over 20 years.

†Based on assumption (Parry, 1962) of 85% of completed manifestation level by age 4½ years

'Proven white' rams and control of clinical scrapie

Fig. A.1 shows the quantitative relationships between the presumptive genotype of the rams siring the females retained over two decades in one high-incidence flock with the scrapie attack-rates observed in each birth cohort. The use of allele-free rams has been accompanied by the disappearance of clinical scrapie. Since the 1968 cohort no case has occurred in the main flock females, although two females retained for ram test-mating and six introduced stud rams have become affected. Similar results have been obtained in two other flocks originally severely affected. In the breeder group, which has been using 'proven white' rams and their sons, continuous data are available for the period 1953–1977 for 13–18 flocks with 1900 to 2900 flock ewes. The attack-rate was 0·4% in 1953–1954 and reached a peak of 3·2% (85 cases) in 1963–1964, since when it has declined gradually to 0·5% in 1972–1973 and is now 0·2%. Such an attack-rate is not of commercial importance, and is confined to flock sectors of known allele-carriers.

Variability of the scrapie attack-rate

The epidemiology of natural scrapie exhibits wide variations in attack-rates in different flocks in the same years and in the same flock in different years.

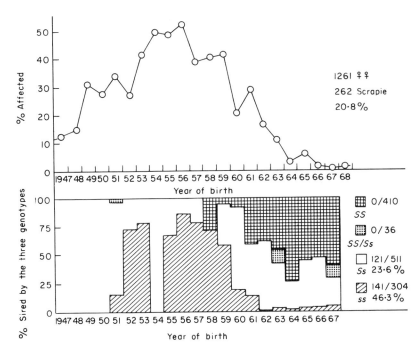

Fig. A1. *The scrapie attack rate in the females of one flock for the period 1947–67 shown by annual birth cohorts in the curve (above) and according to the scrapie genotypes of their sires in the histogram (below). (Reprinted by permission from* Nature, *vol. 277, no. 5692, p. 128. (c) 1979 Macmillan Journals Ltd.)*

Table A.2 provides an analysis of one flock relevant to this phenomenon. In section A progeny of affected rams born in two distinct periods when the flock ewe allele-frequencies varied widely are assembled by dam genotype categories. In section B progeny born in 2 years of affected rams and one unaffected (subsequently established as a 'proven white') ram show that some rams leave no affected progeny irrespective of the dams' genotype and of a very high incidence environment.

The histopathological assessments (footnotes §, ‖ and ¶) confirm the clinical data, but also demonstrate that the onset of degenerative changes, even those of normal senility, are inhibited if either parent is a non-allele carrier.

Anomalous manifestation

Three 'grey' rams have shown atypical clinical and histopathological features reminiscent of scrapie but sufficiently atypical to be considered non-scrapie by both assessments. Whether these cases should be regarded as

Table A2. *Scrapie among the progeny of affected and unaffected rams born in one flock at different periods over 23 years*

Birth cohorts	A		B	
	1951–1959	1966–1973	1958–1959	
Mean annual flock attack-rate	11·0%	1·4%	14·9%	
Sires	7 affected	8 affected	4 affected	1 allele-free
Total progeny retained and average life span	242/4·7 yrs	103/5·2 yrs	85/4·2 yrs	21/7·5 yrs
Total affected/affected plus unaffected at age 4½ yrs or older	109/195 (56)*	2/71 (3)	45/71 (63)	0/17
Out of 'black' (ss) dams	53/56 (95)	1/2 (50)	21/22 (96)	0/10‡, ||
Out of 'grey' (sS) dams	42/86 (49)	1/25† (4)	18/35 (51)	0/7†, ¶
Out of 'grey'/'grey' dams	14/39 (36)	—	6/12 (50)	
Out of 'white' (SS) dams	0/14§ (0)	0/44 (0)	0/2 (0)	

* % age affected with scrapie.
† Under-manifestation.
‡ Two in each group left affected progeny in the F_2 generation.
§ Nine brains examined histologically and found free of lesions at 9½–10 years old.
|| Nine brains examined histologically and found free of lesions at 7–10½ years old.
¶ Six brains examined histologically and found free of lesions at 7–10½ years old.

an unusual manifestation in the heterozygote or an aetiologically different disorder remains undecided.

Under-manifestation, characterised by an older age of onset and a prolonged disease progression of up to 1–2 years has been encountered, especially in flocks with FW blood lines (see Table A.2, footnote †). The recognition of histopathologically positive scrapie among clinically un-affected aged animals suggests a delayed rate of development of the neural dystrophy.

Similarly a small number of clinically unaffected 'black' animals dying suddenly either before or after 4½ years old have been histologically scrapie-positive, as have other 'blacks' presenting as palsy, dementia, loss of vision, epilepsy and wasting.

Discussion

The goodness-of-fit of these data provides substantial support for the view that the clinical manifestation of scrapie is controlled by genetic factors, its occurrence being determined principally by a single autosomal recessive gene, but with age of onset and rate of progression being influenced by the genome, either directly or through unidentified environmental factors. The cytopathological lesions underlying the clinical manifestation affect preferentially certain neuro-anatomical 'systems' (23) with neurones exhibiting a well-developed sulphur-protein metabolism with neurophysin-type components (232). Studies of known genotypes by light microscopy (23), electron microscopy (39, 40) and immunohistochemistry (231, 232) indicate that the primary disturbance probably affects synaptic terminals with an inability to dispose of unwanted metabolic products, followed by retrograde degeneration, cellular vacuolation and karyorrhexis of the neuronal cell body. This dystrophic process, which affects neurones at risk sequentially, may be well established by age 8 months (231), i.e. 1–3 years before expected clinical manifestation, and is also associated with sudden deaths of apparently unaffected 'black' genotypes in early adult life and in old age.

Scrapie may thus be considered an abiotrophy, i.e. a premature metabolic dystrophy, leading to death and sequential disappearance of specialised neurones, occurring over part or all of the normal life-span of 10 ± 2 years, which may be due to the metabolic activity of the recessive gene or to an infectious communicable 'slow virus', the host's response to which is closely related to the gene. A neurocytopathic factor is present in the brains of affected and unaffected genetically 'black' animals, but not in control 'white' animals of the same flock birth cohorts.

The crucial question is whether the scrapie TSEPA is ever communicable in nature. Cannibalism of placental remains (237) is not sufficiently common to account for more than an occasional occurrence. There is little valid evidence in the absence of any immunological (203) or morphological (76) means for demonstrating the agent. Although the presence of a transmissible agent in sheep affected with natural scrapie is well established (98), nothing is known of its distribution in any natural sheep population or of its aetiological role in the initiation of the natural disease process. Its alleged connection with human disease (124) rests on presumptions for which satisfactory evidence is lacking. The present results have failed to reveal thus far any unequivocal evidence that a naturally communicable agent is involved.

We conclude that in the sheep population studied (and probably in most sheep in Britain): (1) natural scrapie is determined, primarily, by an autosomal recessive gene expressed fully in both sexes; (2) no F_1 progeny of a 'proven' allele-free ram will develop scrapie, whatever the scrapie-status of its dam or of its exposure to cases of the disease; (3) the scrapie attack-rates among the nine progeny groups derived from the reciprocal crosses of the three parental genotypes will conform to the predictions of Mendel's laws; (4) the practical control of the disease, and its elimination if desired, may be achieved by the use of allele-free rams and the retention of their progeny; (5) the dissemination of natural scrapie is not dependent upon a naturally communicable parasitic agent; and (6) the hypothesis most consistent with present evidence is that the scrapie TSEPA is formed *de novo* in each affected animal by the metabolic activity of the natural recessive gene.

I am indebted to Mrs E. Beck and Dr. A. A. Vince for major assistance with the histopathological assessments, and to Dr. G. J. Draper for help with the statistics.

Appendix B

The Nomenclature of Scrapie

The first known account of the disorder comes from England, and dates from about 1730, when it was termed '*rickets*' (167). By the end of the century '*goggles*' or '*rubbers*' were the preferred terms, along with '*shewcroft*' and '*shakers*' (77). '*Scratchie*', '*scrapie*' or '*cuddie trot*' were used in Scotland (204), and the Scottish term '*scrapie*' became generally adopted in Britain in the late 19th century, although shepherds still use colloquially the term '*the plague*', comparable with the German '*die Seuche*' (or disease) (298).

In Germany Leopoldt (181) and Erxleben (115) called the disease '*der Trab*' (the trot). Stumpf (289) speaks of '*das Drehen oder Traben*' (turning or trotting). Other names (4) include '*die Traberseuche*', '*die Gnubberkrankheit*' (nibbling disease), '*die Reiberkrankheit*' (rubbing disease), '*die Schafräude-krankheit*' (sheep mange disease), '*die Wetzkrankheit*' (whetting/honing disease), '*die Zitterkrankheit*' (trembling disease), and '*das Schruckigsein*' (shrugging), but '*die Traberkrankheit*' is the most usual term.

In France no clear nomenclature occurs before about 1810, when Tessier (297), the Superintendent of the Rambouillet merino stud flocks, as reported by Girard of Alfort (129), referred to the disease as '*la maladie convulsive*' (the convulsive disease) or '*la maladie folle*' (the mad disease). The term '*le vertige*' (staggers, or dizziness) had been used by Lamerville (178) for a disorder thought to be scrapie, and indeed *le vertige* is used in French writings as synonomous for *die Traberkrankheit* as late as 1827. Later '*la tremblante*' or '*la prurigo lombaire*' became the usual terms.

In Spain the modern terms are '*la trembladera*' or '*la enfermedad trotoria*'.

In Hungary (4), '*surlókór*' (brushing disease) is the most usual term, with '*ügetö-myavalya*' (trotting disease) for the ataxic form.

In Iceland the condition known as '*rida*' (221) appears to be identical with scrapie.

The Yearbooks of the *United Nations Organisation*, WHO/FAO, (1976–1979) list the disorder under '*scrapie*' with '*tremblante*', '*enfermedad trotoria*', '*prurigo lombaire*' as secondary terms. Their preferred term, in heavy type, is '*PARAPLEXIA ENZOOTICA OVIUM*'!

Bibliography

Author's note: In the compilation of this bibliographical list I have relied heavily on the help of librarians of Oxford's Bodleian Library. For the German references I am under an immense obligation to Dr. H. H. Müller of Berlin, who, following an introduction from Dr. Joan Thirsk, Reader in Agrarian History in Oxford, gave unstinted and most detailed help locating the writings associated with Saxony, the Möglin Agricultural Station in Brandenburg, Silesia and Bohemia, and to Professor K. H. Kaufhold of Göttingen, who has arranged for photocopies of many papers. To them I can only extend my special acknowledgements. For Sweden I have received assistance from Mr. David Philipson of Resö, and for Norway from Dr. M. Warming of Copenhagen. For Hungary I am indebted to Professor T. Szent-Iványi and Dr. P. Gunst of Budapest.

[Editor's note: Items marked with an asterisk are of two kinds: (1) documents to which access is difficult, of which there are photocopies in the archives of the Institute of Agricultural History in Reading; and (2) letters to the author from various sources, which are stored in the same place. They may be consulted by arrangement with the archivist, Institute of Agricultural History, University of Reading, Whiteknights, Reading, Berkshire RG6 2AG (Tel. 0734 875123, extension 475). A photocopying service is available.]

1. Abilgaard, P. C. and Viborg, E. (1800). *Veiledning til en forbedret Faarveavl og de spanske Faars behandling i Danmark og Norge* (transl: Advice for improvement in sheep breeding and management of Spanish sheep in Denmark and Norway). H. Tikjøbs Forlag, Copenhagen.

2. Adams, D. H. and Field, E. J. (1968). The infective process in scrapie. *Lancet* **2**, 714–716.

3. Adams, R. D. (1975). *Diseases of Muscle: a study in pathology*. 3rd ed. Harper and Row, Hagerstown, Maryland.

*4. Áldásy, P. and Süveges, T. (1964). A juhok súrlókórjának etc. (transl: Occurrence of scrapie in Hungarian sheep). (In Hungarian, with English Summary) *Magyar Allatorvosok Lapja* **19**, 463–5.

5. Amyx, H. L., Gibbs, C. J., Gajdusek, D. C. and Greer, W. E. (1981). Absence of vertical transmission of subacute spongiform viral encephalopathies in experimental primates. *Proc. Soc. exp. Biol. Med.* **166**, 469–471.

6. Angyalffy, M. A. (1817). *Grundsatze der Schafkultur*. (Quoted by Karasszon, 1968.)

*7. Anon. Breslau (1828). Entwurf einer kurzen Uebersicht der sogenannten Traberkrankheit der Schaafe als epizootisches Erbübel. Wilhelm Gottlieb Korn, Breslau.

8. Anon. Madrid (1609). *Libro de las Leyes del Consejo de la Mesta*, Madrid (quoted by Ford, 1878).
9. Anon. Pest (1843). Juhászat (Transl. Sheep Breeding). Pest; Beimel (in Hungarian). 'Trotting illness' described on pp. 28–30.
*10. Anon. Wiltshire (1788). On the disease called the Goggles in Sheep; by a Gentleman in Wiltshire. *Bath Papers* **1**, 42–44.
11. Armitage, P. L. and Goodall, J. A. (1977). Medieval Horned and Polled Sheep: The archaeological and inconographic evidence. *Antiquaries Journal* **57**, 73–89.
12. Atkeson, F. W., Heman, L. I. and Eldridge, F. (1944). Inheritance of epileptic type character in brown Swiss cattle. *J. Hered.* **35**, 45–48.
13. Austin, H. B. (1944). *The merino; past, present and probable.* Grahame Book Co., Sydney.
14. B.M.J. (1974). (Leading article), Abiotrophies. *Br. Med. J.* **1**, 337–338.
15. Bakewell, R. (1808). Observations on the influence of soil and climate upon wool. London. (Reference from Carter, 1964.)
16. Bannister, W. (*c* 1800). *Cursory observations on the art of rural economy.* (Copy of book in Bodleian Library.)
17. Barnett, K. C. and Palmer, A. C. (1971). Retinopathy in sheep affected with natural scrapie. *Res. Vet. Sci.* **12**, 383–385.
18. Bearn, A. G. (1972). Wilson's disease. *In: The Metabolic Basis of Inherited Disease.* (J. B. Stanbury, J. B. Wyngaarden and D. S. Fredrickson, Eds), 3rd. edn. Ch. 43, pp. 1033–1050. McGraw-Hill, New York.
19. Beck, E. (1969). Discussion of kuru and scrapie. *In: Virus Diseases and the Nervous System.* (C. W. M. Whitty, J. T. Hughes and F. O. MacCallum, Eds), pp. 140–144. Blackwell Scientific, Oxford.
20. Beck, E. and Daniel, P. M. (1959). Some changes in the hypothalamus and proximal pituitary stalk after stalk section. *J. Physiol., London* **146**, 22–24.
21. Beck, E., Daniel, P. M., Gajdusek, D. C. and Gibbs, C. J. (1969). Similarities and differences in the pattern of pathological changes in scrapie, kuru, experimental kuru and subacute presenile polioencephalopathy. *In: Virus Diseases and the Nervous System* (C. W. M. Whitty, J. T. Hughes and F. O. MacCallum, Eds), pp. 107–120. Blackwell Scientific, Oxford.
22. Beck, E., Daniel, P. M. and Parry, H. B. (1961). Hypothalamic and cerebellar system degeneration in sheep with scrapie. Proc. 4th Int. Cong. Neuropath., Munich. Vol. III, pp. 269–274. Thieme, Stuttgart.
23. Beck, E., Daniel, P. M. and Parry, H. B. (1964). Degeneration of the cerebellar and hypothalamo-neurohypophysial systems in sheep with scrapie; and its relationship to human system degenerations. *Brain* **87**, 153–176.
24. Bedford, Duke of (1795). *An experiment, comparing four breeds of sheep.* Annals Agriculture **23**, 456–467.
25. Bednařík, K. and Havelková, V. (1973–1974). *Oekonomische Neuigkeiten und Verhandlungen:* bibliography. 1; 1811–1832. 2; 1833–1850. Prague.
26. Behrens, H. (1957). Discussion on paper by Palmer: on Scrapie in Germany. *Vet. Rec.* **69**, 1328.
27. Behrens, H. (1979). Krankheiten. *In: Lehrbuch der Schafzucht.* (H. Behrens, H. Doehner, R. Scheelje and R. Wassmuth, Eds), Chap. 7, pp. 226–319. (*Traberkrankheit* discussed on p. 232.) Paul Parey, Hamburg.
*27a. Belda, A. S. (1979). Letter to H. B. Parry dated 9.7.79. Jefe de la sección de Ganado Bovino. Ministerio de Agricultura, Madrid.

28. Belda, A. and Trujillano, M. C. S. (1979). *Las razas ovinas españolas* (Publicaciones de Extensión Agraria, Bravo Murillo, 101, Madrid).

29. Belschner, H. G. (1959). *Sheep management and diseases.* 6th ed., Angus and Robertson, Sydney.

30. Bénion, A. (1874). Traité complet de l'élevage et des maladies du mouton. Anselin, Paris.

*31. Berger, M. (1829). Observations de paraplégie sur deux moutons. *Rec. méd. vét.* **6**, 615–619.

32. Bergland, R. M. and Torack, R. M. (1969). An electron microscopic study of the human infundibulum. *Z. Zellforsch.* **99**, 1–12.

*33. Bertrand, J., Carré, H. and Lucam, F. (1937). La 'tremblante' du mouton (Recherches histo-pathogiques). *Ann. Anat. path.* **14**, 565–586.

34. Besnoit and Morel, C. (1898). *Notes sur les lesions nerveuses de la tremblante du mouton. C.R. Soc. Biol.* **5**, 536.

*35. Bewick, R. (1791). *A general history of quadrupeds* with figures engraved on wood by T. Bewick, 2nd edn. Newcastle-on-Tyne and London.

*36. Bewick, T. (1807). *A general history of quadrupeds.* 5th edn. Newcastle-upon-Tyne, printed by Edward Walker for T. Bewick and S. Hodgson.

37. Bignami, A., Beck, E. and Parry, H. B. (1970). Neurosecretion-like material in the hindbrain of aging sheep and sheep affected with natural scrapie. *Nature (London)* **225**, 194–196.

38. Bignami, A. and Parry, H. B. (1971). Aggregation of 35-nanometer particles associated with neuronal cytopathic changes in natural scrapie. *Science* **171**, 389–390.

39. Bignami, A. and Parry, H. B. (1972a). Electron microscopic studies of the brain of sheep with natural scrapie. I. The fine structure of neuronal vacuolation. *Brain* **95**, 319–326.

40. Bignami, A. and Parry, H. B. (1972b). Electron microscopic studies of the brain of sheep with natural scrapie. II. The small nerve processes in neuronal degeneration. *Brain* **95**, 487–494.

41. Billingsley, J. (1798). *General View of the Agriculture of the County of Somerset.* 2nd edn. London. (See also *Bath papers*, 1792, **7**).

42. Bonduelle, M., Escourolle, R., Bouygues, P., Lormeau, G., Dumas, J-L. R. and Merland, J. J. (1971). Maladie de Creutzfeldt-Jakob familiale; observation anatomo-clinique. *Rev. Neurol.* **125**, 197–209.

43. Bosanquet, F. D., Daniel, P. M. and Parry, H. B. (1956). Myopathy in sheep; its relationship to scrapie and to dermatomyositis and muscular dystrophy. *Lancet* **2**, 737–746.

44. Brash, A. G. (1952a). First outbreak of scrapie reported in New Zealand. *N.Z. J. Agric.* **85**, 305–306.

45. Brash, A. G. (1952b). Scrapie in imported sheep in New Zealand. *N.Z. Vet. J.* **1**, 27–30.

*46. Braudel, F. (1972). *The Mediterranean and the Mediterranean World in the Age of Philip II.* Vol. I (Transl. from the French by Sian Reynolds). pp. 91–93. Harper and Row, New York.

47. Brock, D. J. H. and Mayo, O. Eds (1978). *The Biochemical Genetics of Man* (2nd edn.). Academic Press, London.

48. Brodal, A. (1939). Experimentelle Untersuchungen über retrograde Zellveränderungen in der unteren Olive nach Läsionen des Kleinhirns. *Z. ges. Neurol. Psychiat.* **166**, 624–704.

49. Brodal, A. (1957). *The reticular formation of the brainstem. Anatomical aspects and functional correlations.* Oliver and Boyd, Edinburgh.
50. Brodal, A. and Gogstad, A. C. (1954). Rubro-cerebellar connection. An experimental study in the cat. *Anat. Rec.* **118**, 455–486.
51. Brodal, A., Pompeiano, O. and Walberg, F. (1962). *The vestibular nuclei and their connections, anatomy and functional correlations.* Oliver and Boyd, Edinburgh.
52. Brotherston, J. G., Renwick, C. C., Stamp, J. T., Zlotnik, I. and Pattison, I. H. (1968). Spread of scrapie by contact to goats and sheep. *J. Comp. Path.* **78**, 9–17.
53. Brownlee, A. (1940). Histopathological studies of scrapie, an obscure disease of sheep. *Vet. J.* **96**, 254–264.
54. Bruce, M. E. and Fraser, H. (1975). Amyloid plaques in the brains of mice infected with scrapie: morphological variation and staining properties. *Neuropath. appl. Neurobiol.* **1**, 189–202.
55. Bruere, A. N. (1977). Scrapie: a point of view. *N. Zealand Vet. J.* **25**, 259–260.
56. Bull, L. B. and Murnane, D. (1958). An outbreak of scrapie in British sheep imported into Victoria. *Austral. Vet. J.* **34**, 213–215.
57. Bunn, H. F., Forget, B. G. and Ranney, H. M. (1977). *Human Hemoglobins.* pp. 108–112. W. B. Saunders, Philadelphia.
58. Buntain, D., Heath, G. B. and Thompson, J. R. (1974). Scrapie: observations on a field outbreak. *Vet. Rec.* **92**, 332.
59. Burch, P. R. J. (1968). *An Inquiry Concerning Growth, Disease and Ageing.* Oliver and Boyd, Edinburgh.
60. Burger, D. and Hartsough, G. R. (1965). Encephalopathy of mink: II. Experimental and natural transmission. *J. Infect. disease* **115**, 393–399.
61. Cadiot, P.–J., Lesbouyries, G. and Ries, J.-N. (1925). *Traité de médecine des animaux domestiques,* pp. 531–532. Vigot, Paris.
*62. Caggiano, C. H. (1977). Letter to H. B. Parry, dated 15.6.77. Director General, Animal Health Service, Buenos Aires.
63. Cameron, J. (1913). Lecture to Northumberland Shepherds' Society at Ancroft, Northumberland, February 1913. Berwickshire News, February 18th, p. 6.
64. Carlier, C. (1769). *Traité des bêtes à laine.* Paris.
65. Carman, E. A. (1891). The Sheep of Great Britain. Ann. Rep. Bur. Anim. Industry, U.S. Dept. Agric., Washington, D.C., pp. 145–246.
66. Carr, R. (1966). *Spain: 1808–1939.* Clarendon Press, Oxford.
67. Carré, H. (1923). Sur la tremblante du mouton. *Rev. gén. Méd. vét.* **32**, 239–241.
68. Carter, H. B. (1964). *His Majesty's Spanish flock. Sir Joseph Banks and the merinos of George III of England.* Angus and Robertson, London.
69. Cassirer, R. (1898). Ueber die Traberkrankheit des Schafe. Pathologisch-anatomische und bakterielle Untersuchung. *Virchows Arch. Path. Anat. Physiol,* **153**, 89–110.
70. Cauvet, J. (1854). Sur la tremblante. *J. des Vétérinaires du midi* **7**, 441–448.
71. Cavanagh, J. B. (1964). The significance of the 'dying-back' process in experimental and human neurological disease. *Int. rev. Exper. Path.* **7**, 219–267.
72. Cavanagh, J. B. (1979). The 'dying-back' process: a common denominator in many naturally occurring and toxic neuropathies. *Arch. Pathol. Lab. Med.* **103**, 659–664.
73. Chandler, R. L. (1961). Encephalopathy in mice produced by inoculation with scrapie brain material. *Lancet* **1**, 1378–1379.
74. Chandler, R. L. and Fisher, J. (1963). Experimental transmission of scrapie to rats. *Lancet* **2**, 1165.

75. Chelle, P. L. (1942). Un cas de tremblante chez la chèvre. *Bull. Acad. vét. Fr.* **15**, 294–295.

76. Cho, H. G., Greig, A. S., Corp, C. R., Kimberlin, R. H., Chandler, R. L. and Millson, G. C. (1977). Virus-like particles from both control and scrapie-affected mouse brain. *Nature (London)* **267**, 459–60.

*77. Claridge, J. (1795). Extract from a General View of Agriculture, in the County of Dorset; with observations on the means of its improvement. *Bath Papers* **7**, 66–73.

78. Cleeve, H. (1840). Practical Essay on Diseases of Sheep. *J. Eng. Agric. Soc.* **1**, 295–345.

79. Cobb, W. A. (1975). Electroencephalographic changes in viral encephalitis. Chapter 5 in: (L. S. Illis, Ed.) *Viral Diseases of the Central Nervous System.* pp. 76–89, Baillière Tindall, London.

80. Cobb, W. A., Hornabrook, R. W. and Sanders, S. (1973). The EEG of Kuru. *Electroenceph. clin. Neurophysiol.* **34**, p. 419–427.

81. Collins, J. (1799). Some further practical remarks on the nature of sheep and wool; and the disorders of sheep etc. *Bath Papers* **9**, 113–128.

*82. Comber, T. (1772). Real improvements in agriculture (on the principles of A. Young, Esq.). Letters to Reade Peacock, Esq. and to Dr. Hunter, Physician in York, concerning the Rickets in sheep. Nicoll, London.

83. Cooper, J. E. (1973). A report of scrapie in sheep in Kenya. *Br. Vet. J.* **129**, 13–16.

84. Court, L., Cathala, F., Bouchard, N., Breton, P. and Gourmelon, P. (1979). Electrophysiological and behavioural findings in the natural and experimental spongiform encepalopathies. *In: Slow Transmissible Diseases of the Nervous System* (Prusiner, S. B. and Hadlow, W. J., Eds), Vol. **1**, pp. 305–328. Academic Press, New York.

85. Cox, E. W. (1936). *The Evolution of the Australian Merino.* Sydney.

86. Cragg, B. G. (1975). The density of synapses and neurones in normal, mentally defective and aging human brains. *Brain* **98**, 81–90.

87. Cuillé, J. and Chelle, P. L. (1936). La maladie dite tremblante du mouton est-elle inoculable? *C. R. Acad. Sci. Paris* **203**, 1552–1554.

88. Culley, G. (1807). Observations on Livestock, containing hints for choosing and improving the best breeds of the most useful domestic animals. London.

88a. Daniel, P. M. (1971). Transmissible degenerative diseases of the nervous system. *Proc. Roy. Soc. Med.* **64**, pp. 787–794.

89. Daniel, P. M. and Prichard, M. M. J. (1957a). The vascular arrangements of the pituitary gland of the sheep. *Quart. J. exp. Physiol.* **42**, 237–247.

90. Daniel, P. M. and Prichard, M. M. J. (1957b). Anterior pituitary necrosis in the sheep produced by section of the pituitary stalk. *Quart. J. Exp. Physiol.* **42**, 248–253.

91. Daniel, P. M. and Treip, C. S. (1976). The Hypothalamus and Pituitary Gland. *In: Greenfield's Neuropathology*, 3rd edn. (W. Blackwood and J. A. N. Corsellis, Eds), Ch. 13, pp. 581–607. Arnold, London.

92. Daubenton, L.-J.-M. (1782). *Instructions pour les Bergers et pour les Propriétaires des Troupeaux.* Paris.

*93. Daubenton, L.-J.-M. (1800). (English Translation.) Instructions for shepherds and owners of flocks. Extracts from the foreign communications to the Board of Agriculture; pp. 223–242; 261–275.

94. Davis, T. (1795). Extract from a general view of the Agriculture of the County of Wiltshire; with observations on the Means of its Improvement; drawn up for the Board of Agriculture and Internal Improvement. *Bath Papers* **7**, 113–221.

*95. Davis, T. (1811). General View of the Agriculture of Wiltshire. pp. 140–149. Richard Phillips, London.

96. Dellman, H. D. and Rodriguez, E. M. (1970). Herring Bodies; an electron microscopic study of local degeneration and regeneration in neurosecretory axons. *Z. Zellforsch.* **111**, 293–315.

*97. de Poederlé, Baron (1797). Answers to queries by Sir John Sinclair. Communications to the Board of Agriculture on subjects relative to the husbandry, etc. Vol. 1, pp. 247–255. London.

98. Dickinson, A. G. (1976). Scrapie in sheep and goats. *In: Slow Virus Diseases of Animals and Men* (R. H. Kimberlin, Ed.), Ch. 10. North Holland Publishing Company, Amsterdam.

99. Dickinson, A. G. and Fraser, H. (1979). An assessment of the genetics of scrapie in sheep and mice. *In: Slow Transmissible Diseases of the Nervous System* (S. B. Prusiner and W. J. Hadlow, Eds), Vol. I, pp. 367–385. Academic Press, New York.

100. Dickinson, A. G., Stamp, J. T. and Renwick, C. C. (1974). Maternal and lateral transmission of scrapie in sheep. *J. Comp. Pathol.* **84**, 19–25.

101. Dickinson, A. G., Young, G. B., Stamp, J. T. and Renwick, C. C. (1965). An analysis of natural scrapie in Suffolk sheep. *Heredity* **20**, 485–503.

102. Diepen, R. (1962). The difference in the neurosecretory picture in various mammals. *In: Neurosecretion* (H. Heller and R. B. Clark, Eds), pp. 111–121. Academic Press, London and New York.

103. Dixon, H. H. (1869). Report on Farms. *J. Roy. Agr. Soc. England.* **5**(*ii*), 394–395.

104. Doehner, H. (1944). *Handbuch der Schafzucht und Schafhaltung.* Vol. 3. p. 609. Paul Parey, Berlin.

*105. Downie, C. (1811). Particulars regarding the Merino sheep, imported by Charles Downie, Esq. Communications to the Board of Agriculture on subjects relative to the husbandry, etc. Vol. 7, part 1, pp. 61–63.

106. Draper G. J. (1963). 'Epidemics' caused by a late-manifesting gene. Application to scrapie. *Heredity* **18**, 165–171.

107. Draper, G. J. and Parry, H. B. (1962). Scrapie in sheep: the hereditary component in a high incidence environment. *Nature (London)* **195**, 670–672.

108. Duffy, P., Wolf, J., Collins, G., DeVoe, A. G., Streeten, B and Cowen, D. (1974). Possible person-to-person transmission of Creutzfeldt-Jakob disease (letter), *New England J. Med.* **290**, 692–693.

109. Eisenmáyer, S. and Göbel, C. (1872). *The healing of animals*, etc. Pest. (In Hungarian.)

110. Ellis, J. T. (1956). Necrosis and regeneration of muscle in cortisone-treated rabbits, *Am. J. Pathol.* **32**, 993–1013.

111. Ellman, J. (1793). On Sheep. *Annals Agriculture* **20**, 172–179.

112. Epstein, H. (1971). The origin of the domestic animals of Africa. Vol. 2. Africana Publishing Co., New York.

113. Erdt, W. E. A. (1861). *Die Traberkrankheit der Schafe.* Bosselmann, Berlin.

114. Erickson, R. P. (1972). Leber's optic atrophy, a possible example of maternal inheritance. Annotation. *Am. J. Human Genetics* **24**, 348–349.

115. Erxleben, J. C. P. (1769). Betrachten über das Studium der Vieharzneykunst. Göttingen. (Dissertation.)

116. Farrant, S. (1978). John Ellman of Glynde in Sussex. *Agric. Hist. Rev.* **26**, part II, 77–88.

117. Field, E. J. and Shenton, B. K. (1973). Rapid immunological method for diagnosis of natural scrapie in sheep. *Nature (London)* **244**, 96–97.

*118. Fink, J. H. (1797). Answers to queries concerning the breeding of sheep in Germany, particularly Upper Saxony. Communications to the Board of Agriculture (UK) on subjects relative to the husbandry, etc. Vol. I, pp. 276–294. London.

119. Ford, R. (1846). *Gatherings from Spain.* Murray's Travellers Series. John Murray, London.

120. Ford, R. (1878). *A Handbook for Travellers in Spain.* 5th edn. John Murray, London.

121. Fraser, H. and Hancock, P. M. (1977). An investigation of the macrophage electrophoretic mobility test in the diagnosis of scrapie in sheep. *J. Comp. Pathol.* **87**, 267–274.

122. Fribourg, A. (1910). La transhumance en Espagne. *Annales de Géographie* **19**, 231–244.

123. Gaiger, S. H. (1924). Scrapie. *J. Comp. Path. Ther.* **37**, 259–277.

124. Gajdusek, D. C. (1977). Unconventional viruses and the origin and disappearance of Kuru. *Science* **197**, 943–960.

125. Gajdusek, D. C. and Zigas, V. (1957). Degenerative disease of the central nervous system in New Guinea. The endemic occurrence of 'Kuru' in the native population. *New Eng. J. Med.* **257**, 974–978.

126. Garrod, A. E. (1909). *Inborn Errors of Metabolism.* Hodder and Stoughton, London. (Reprinted with supplement by H. Harris, 1963, by Oxford University Press, Oxford).

127. Garzuly, F., Jellinger, K. and Pilz, P. (1971). Subakute spongiöse Encephalopathie (Jakob-Creutzfeldt Syndrom). Klinisch-morphologische Analyse von 9 Fällen. *Arch. Psychiatr. Nervenkr.* **214**, 207–227.

128. Gibbons, R. A. and Hunter, G. D. (1967). Nature of the Scrapie Agent. *Nature (London)* **215**, 1041–1043.

*129. Girard, J. (1829–30). Notice sur quelques maladies peu connues des betes à laine. *Rec. Méd. Vét.* **6**, 674–683; **7**, 26–39 and 65–76.

*130. Golf, A. (1939). Ueber die frühere Einführung der Merinos aus Spanien nach Deutschland und anderen Ländern. *In: Handbuch der Schafzucht und Schafhaltung* (H. Doehner, Ed.), Vol. 1, Ch. 3, pp. 63–72. Paul Parey, Berlin.

131. Gooch, W. (1811). General view of the Agriculture of the County of Cambridge. Sheep, pp. 272–279. London.

132. Gordon, W. S. (1946). Advances in veterinary research. *Vet. Rec.* **58** (47), 516–525.

133. Gowers, W. R. (1902). A lecture on abiotrophy. *Lancet* **1**, 1003–1007.

134. Greenfield, J. G. (1954). *The spino-cerebellar degenerations.* Blackwell, Oxford.

135. Greig, J. R. (1940). Scrapie. *Trans. Highland and Agric. Soc., Scotland* **52**, 71–90.

136. Guillemin, R. (1972). Characterization and Identification of Pituitary Releasing Factors from the Hypothalamus. *Neurosci. Res. Prog. Bull.* **10**, 193–199.

*137. Gunst, P. (1979a). Letter to H. B. Parry, dated 19.4.79. (Keleman Laszlo V.9; Budapest.)

*138. Gunst, P. (1979b). Letter to Dr. Joan Thirsk, dated 15.11.79.

139. Guthrie, J. F. (1957). *A World History of Sheep and Wool.* Pastoral Review Pty. Ltd, Melbourne.

140. Halász, B. (1969). The endocrine effects of isolation of the hypothalamus from the rest of the brain. *In: Frontiers in Neuroendocrinology* (W. F. Ganong and L. Martini, Eds), pp. 307–342. Oxford University Press, Oxford.

141. Haltia, M., Kovanen, J., Van Crevel, H., Bots, G. T. A. M. and Stefanko, S. (1979). Familial Creutzfeldt-Jakob disease. *J. Neurol. Sci.* **42**, 381–389.

142. Haralambiev, H., Ivanov, I., Vesselinova, A and Mermerski, K. (1973). An attempt to induce scrapie in local sheep in Bulgaria. *Zbl. Vet. Med. B.* **20**, 701–709.

143. Hartley, D. (1979). *The Land of England.* Macdonald, London.

144. Hartsough, G. R. and Burger, D. (1965). Encephalopathy of Mink: I Epizootiologic and clinical observations. *J. Infect. Dis.* **115**, 387–392.

145. Herzberg, L., Herzberg, B. N., Gibbs, C. J. J., Sullivan, W., Amyx, H. and Gajdusek, D. C. (1974). Creutzfeldt-Jakob disease: hypothesis for high incidence in Libyan Jews in Israel. *Science* **186**, 848.

146. Hicks, S. P. and Coy, M. A. (1958). Pathologic effects of antimetabolites. Convulsions and brain lesions caused by sulfoximine and their variation with genotype. *Arch. Path.* **65**, 378–387.

147. Hiepe, von Th., Bergmann, V. and Jungmann, R. (1973). Beitrag zur klinischen und pathomorphologischen Diagnostik der Traberkrankheit des Schafes. *Monatsschr. Vet. Med.* **28**, 905–907.

148. Highland Society (1789). Report of the Highland Society of Scotland on the subject of Shetland Wool. Edinburgh.

149. Hild, W. and Zetler, G. (1953). Experimenteller Beweis für die Entstehung der sog. Hypophysenhinterlappenwirkstoffe im Hypothalamus. *Pflügers Arch. ges. Physiol.* **257**, 169–201.

150. Hoare, M., Davies, D. C. and Pattison, I. H. (1977). Experimental production of scrapie-resistant Swaledale sheep. *Vet. Rec.* **101**, 482–484.

151. Holman, H. H. and Pattison, I. H. (1943). Further evidence on the significance of vacuolated nerve cells in the medulla oblongata of sheep affected with scrapie. *J. Comp. Path.* **53**, 231–236.

152. Holmes, G. and Stewart, T. G. (1908). On the connection of the inferior olives with the cerebellum in man. *Brain* **31**, 125–137.

153. Hoorens, J. and Oyaert, W. (1966). Scrapie in Sheep. *Vlaams Diergeneeskundig Tijdschr.* **35**, 313–317.

154. Hope, D. B. and Pickup, J. C. (1974). Neurophysins. *In: Handbook of Physiology.* Section 7, Endocrinology, Vol. 4. Pituitary gland, pt. 1. American Physiological Soc., Washington.

*155. Hourrigan, J. L. (1979). Experience with natural scrapie in sheep and goats at Mission, Texas. Paper presented at the Scrapie Conference at College Park, Maryland in March 1977.

156. Hourrigan, J., Klingsporn, A., Clark, W. W. and de Camp, M. (1979). Epidemiology of scrapie in the United States. *In: Slow Transmissible Diseases of the Nervous System* (S. B. Prusiner and W. J. Hadlow, Eds), Vol. I, pp. 331–356. Academic Press, New York.

157. Hulland, T. J. (1958). The skeletal muscle of sheep affected with scrapie. *J. Comp. Path. Therap.* **68**, 264–274.

158. Humfrey, W. (1840). Footnote to Cleeve's Practical essay in diseases of sheep. *J. Eng. Agric. Soc.* **1**, 297.

159. Hunter, G. D. (1972). Scrapie: a prototype slow infection. *J. Inf. Dis.* **125,** 427–440.

160. Hutyra, F. (1888). *Kórboneztani diagnosztika* . . . (Introduction to the morbid anatomy of domestic animals: in Hungarian). *Athenaeum R. Tars.,* Budapest.

*161. Ivanov, I. and Haralambiev, K. (1970). Scrapie in sheep in Bulgaria and problems associated with it. *Vet. Med. Nanki, Sofia* **7,** 11.20. (In Bulgarian.)

162. Jacob, H. (1957). Sekundäre, retrograde und transsynaptische Degeneration. *In: Handbuch der speziellen pathologischen Anatomie und Histologie.* (O. Lubarsch, F. Henke and R. Rössle, Eds), Vol. XIII, 1A, pp. 266–336.

163. Jacobeit, W. (1961). *Schafhaltung and Schafer in Zentraleuropa bis zum Beginn des 20. Jahrhunderts,* Berlin. (Quoted by Braudel, 1972).

164. Jinks, J. L. (1964). *Extrachromosomal Inheritance.* Prentice-Hall, New Jersey.

165. Joubert, L., Lapras, M., Gastellu, J., Prave, M. and Laurent, D. (1972). Un foyer de tremblante du mouton en Provence. *Bull. Soc. Sci. Vét. et Méd. comparée, Lyon* **74,** pp. 165–184.

166. Joubert, L., Bonneau, M. and Carrier, H. (1974). La tremblante, maladie lente a viroïde. Pathogénie, épidémiologie, prophylaxie. *Rev. Méd. Vét.* **125,** 647–666.

167. Journal of the House of Commons (1755). Vol. 27, pp. 87.

168. Kahana, A., Alter, M., Braham, J. and Sofer, D. (1974). Creutzfeldt-Jakob disease. Focus among Libyan Jews in Israel. *Science* **183,** 90–91.

169. Karasszon, D. (1968). A history of scrapie in sheep. *Magyar Allatorvosok Lapja* **23,** 383–384 (in Hungarian).

*170. Katiyar, R. D. (1961). Letter to H. B. Parry, dated August 1961. [Assistant Disease Investigation Officer (Sheep and Goats) Pashulok, India.]

*170a. Katiyar, R. D. (1962). A preliminary report on the occurrence of scrapie in Indian sheep. *Ceylon Vet. J.* **10,** 93–96.

171. Kimberlin, R. H. (1976) (Ed.). *Slow virus diseases of animals and man.* North Holland/American Elsevier, Amsterdam.

172. Kimberlin, R. H. (1979). Aetiology and genetic control of natural scrapie. *Nature (London)* **278,** 303–304.

173. Klein, J. (1920). *The Mesta; a study in Spanish Economic History, 1273–1836.* Harvard University Press, Cambridge, Mass. (Harvard Economic Studies, Vol. 21).

174. Knowles, D. (1950). *The religious orders in England.* Cambridge University Press, Cambridge.

175. Krücke, W., Beck, E. and Vitzthum, H. (1973). Creutzfeldt-Jakob disease. Some unusual morphological features reminiscent of Kuru. *Z. Neurol.* **206,** 1–24.

176. Kurtzke, J. F. (1977). Multiple sclerosis from an epidemiological viewpoint. *In: Multiple Sclerosis: a Critical Conspectus* (E. J. Field, Ed.), Ch. 4, pp. 83–142. MTP Press, Lancaster.

177. Ladurie, E. (1971). *Times of feast, times of famine: a history of climate since the year 1000.* Doubleday, London.

178. Lamerville, M. le Chev. (1788). Review of *Observations pratiques sur les bêtes à laine. Ann. Agric.* **9,** 174–178. Buisson, Paris.

179. Larsell, O. (1970). The comparative anatomy and histology of the cerebellum. *In: Monotremes through apes.* (J. Jansen, Ed.). University of Minnesota Press, Minnesota.

180. Lasteyrie, C. P. (1810). An account of the introduction of Merino Sheep into the different States of Europe and at the Cape of Good Hope. London (Translated from the French).

*181. Leopoldt, J. G. (1759). *Nützliche und auf die Erfahrung gegründete Einleitung zur Landwirtschaft.* Glogau, Berlin.

182. Littlejohn, A. I. and Terlecki, S. (1967). Scrapie in a Dorset Horn Ewe. *Vet. Rec.* **81**, 355–356.

183. Livett, B. G. (1975). Immunochemical studies on the storage and axonal transport of neurophysins in the hypothalamo-neurohypophysial system. *Ann. N. Y. Acad. Sci.* **248**, 112–133.

184. Livett, B. G. and Parry, H. B. (1971). Accumulation of neurophysin in the median eminence and the cerebellum of sheep with natural scrapie. *Br. J. Pharmacol.* **43**, 423–424.

185. Livett, B. G. and Parry, H. B. (1972). The distribution of vasopressin and neurophysin in the hypothalamus-distal-neurohypophysial and hypo-thalamo-infundibular neurosecretory systems of normal and scrapie-affected sheep. *J. Physiol.* **230**, 20–22.

186. Livett, B. G., Uttenthal, L. O. and Hope, D. B. (1971). Localisation of neuro-physin II in the hypothalamo-neurohypophysial system of the pig by immuno-fluorescence histochemistry. *Phil. Trans. Roy. Soc. London* B, **261**, 371–378.

187. Llinás, R. (1975). The cortex of the cerebellum. *Sci. Am.*, **232**, 56–71.

188. Locke, W. and Schally, A. V. (1972). *The Hypothalamus and Pituitary in Health and Disease.* Charles C. Thomas, Springfield, Illinois.

*189. Lopez, R. S. (1953). The origin of the Merino Sheep. *In: Josiah Starr Memorial Volume:* Jewish Social Studies Publication No. 5, pp. 161–168. New York.

190. Low, D. (1845). *On the Domesticated Animals of the British Islands.* London.

*191. Lucam, F., Béchade, R. and Saurat, P. (1950). La 'Tremblante' du Mouton dans le départment de L'Indre. *Bull. Acad. Vét.* **23**, 321–325.

192. Malingié-Nouel, M. (1853). On a method of obtaining fixity of type in forming a new breed of sheep. *J. Roy. Agr. Soc. Engl.* **14**, 214–224.

193. Marsh, R. F. (1976). The subacute spongiform encephalopathies. *In: Slow Virus Diseases of Animals and Man* (R. H. Kimberlin, Ed.), Ch. 15. Elsevier/North Holland, Amsterdam.

194. Masters, C. L., Gajdusek, D. C. and Gibbs, C. J. (1981). The familial occur-rence of Creutzfeldt-Jakob disease and Alzheimer's disease. *Brain* **104**, 535–558.

195. Masters, C. L., Bernoulli, C. and Asher, D. M. (1979). Familial Creutzfeldt-Jakob disease and other familial dementias: an enquiry into possible modes of transmission of virus-induced familial diseases. *In: Slow Transmissible Diseases of the Nervous System* (S. B. Prusiner and W. Hadlow, Eds), Vol. 1, pp. 143–194. Academic Press, New York.

196. Masters, C. L., Kakulas, B. A., Alpers, M. P., Gajdusek, D. C. and Gibbs, C. J. (1976). Preclinical lesions and their progression in the experimental spongiform encephalopathies (Kuru and Creutzfeldt-Jakob disease) in primates. *J. Neuropath. exp. Neurol.* **35**, 593–605.

197. Matson, G. A., Schut, J. W. and Swanson, J. (1961). Hereditary Ataxia: linkage studies in hereditary ataxia. *Ann. Hum. Genet.* **25**, 7–23.

198. Mauron, M. (1952). La transhumance du pays d'Arles aux grandes Alpes. (Quoted by Braudel, 1972.)

*199. May, G. (1868). *Das Schaf: seine Wolle, Racen, Züchtung, Ernährung und Benut-zung, sowie dessen Krankheiten.* Band 2; *die inneren und äusseren Krankheiten des Schafes,* Trewendt, Breslau.

*200. Mehnert, E. (1979). Letter to H. B. Parry, dated 25.9.79. (Director, Division of Contagious Animal Diseases, National Board of Agriculture, Jonkoping, Finland.)

201. Mitchell, J. (1827). *Sketches of Agriculture or Farmers' Remembrancer, alphabetically arranged.* Baldwin, Cradock and Taylor, London.

202. M'Fadyean, J. (1918). Scrapie. *J. Comp. Path. Ther.* **31**, 102–131.

203. McFarlin, D. E., Raff, M. C., Simpson, E. and Nehlsen, S. H. (1971). Scrapie in immunologically deficient mice. *Nature (London)* **233**, 336.

*204. McGowan, J. P. (1914). *Investigation into the disease of sheep called 'scrapie', with special reference to its association with sarcosporidiosis.*(Rept. 223, Edinburgh and East of Scotland Coll. of Agric. 1914) Blackwood, Edinburgh.

205. McKusick, V. A. (1969). *Human Genetics.* 2nd. edn. Prentice-Hall, Englewood Cliffs, New Jersey.

206. N.S.A. (1979). *British Sheep.* 5th edn., published by the National Sheep Association (UK).

207. Naerland, G. (1970a). Sheep Husbandry in Norway. *Vet. Rec.* **86**, 129–133.

*208. Naerland, G. (1970b). Letter to H. B. Parry, dated 23.6.70. (Veterinaer, Sandnes, Norway.)

*209. Naerland, G. (1979). Letter to H. B. Parry, dated 29.3.79. (Veterinaer, Sandnes, Norway.)

210. Nance, W. E. (1969). Anencephaly and spina bifida: a possible example of cytoplasmic inheritance in man. *Nature, London* **224**, 373–375.

211. Nedkvitne, J. J. (1955). Or soga om merinosauen, i Noreg. *Tidsskrift for Det norske Landbruk* **62**, 20–46.

212. Nussbaum, R. E., Henderson, W. M., Pattison, I. H., Elcock, N.V. and Davies, D. C. (1975). The establishment of sheep flocks of predictable susceptibility to experimental scrapie. *Res. Vet. Sci.* **18**, 49–58.

213. Oppenheimer, D. R. (1975). Pathology of transmissible and degenerative diseases of the nervous system. *In: Viral Diseases of the Central Nervous System* (L. S. Illis, Ed.). Baillière Tindall, London.

214. Oppenheimer, D. R. (1976). Diseases of the Basal Ganglia, Cerebellum and Motor Neurons. *In: Greenfield's Neuropathology* (W. Blackwood and J. A. N. Corsellis, Eds), 3rd edn. Ch. 14, pp. 608–651. Arnold, London.

215. Oppermann (1919). *Lehrbuch der Krankheit des Schafes.* p. 234. Hanover (also 2nd edn. 1921).

216. Orwin, C. S. and Whetham, E. G. (1971). *History of British Agriculture, 1846–1914.* 2nd edn. David and Charles, Newton Abbot.

*217. Øverås, J. (1979). Letter to H. B. Parry, dated 19.6.79. (Statens Veterinaere Forsøksgard for Småfe, Sandnes, Norway.)

218. Palmer, A. C. (1957a). Vacuolated neurones in sheep affected with scrapie. *Nature (London)* **179** 480–481.

219. Palmer, A. C. (1957b). Studies in scrapie. *Vet. Rec.* **69**, 1318–1327.

220. Palmer, A. C. (1958). Anatomical arrangement of the grey matter in the medulla, pons and midbrain of the sheep. *Zbl. f. Veterinärmedizin* **10**, 953–967.

221. Pálsson, P. A. (1979). Rida (Scrapie) in Iceland and its epidemiology. *In: Slow Transmissible Diseases of the Nervous System* (S. B. Prusiner and W. J. Hadlow, Eds), Vol. I, pp. 357–366. Academic Press, New York.

222. Pálsson, P. A. and Sigurdsson, B. (1959). A slow progressive disease affecting the central nervous system of sheep. Proc. VII. Nord. Vet. Congr. Helsinki, 1958, pp. 179–191 (in Danish).

223. Parry, H. B. (1953). Degenerations of the Dog Retina. II. Generalized progressive atrophy of hereditary origin. *Brit. J. Ophth.* **37**, 487–502.

224. Parry, H. B. (1957). Scrapie and related myopathies in sheep. *Vet. Record.* **69**, 43–55.

225. Parry, H. B. (1960). Scrapie: a transmissible hereditary disease of sheep. *Nature (London)* **185**, 441–443.
226. Parry, H. B. (1962). Scrapie: a transmissible and hereditary disease of sheep. *Heredity* **17**, 75–105.
227. Parry, H. B. (1969). Scrapie—natural and experimental. *In: Virus Diseases and the Nervous System* (C. W. M. Whitty, J. T. Hughes and F. O. MacCallum, Eds), pp. 99–105. Blackwell Scientific Press, Oxford.
228. Parry, H. B. (1979a). Elimination of natural scrapie in sheep by sire genotype selection. *Nature (London)* **277**, 127–129.
229. Parry, H. B. (1979b). Aetiology of natural scrapie. *Nature (London)* **280**, 12.
230. Parry, H. B. and Livett, B. G. (1973). A new hypothalamic pathway to the median eminence containing neurophysin and its hypertrophy in sheep with natural scrapie. *Nature (London)* **242**, 63–65.
231. Parry, H. B. and Livett, B. G. (1976). Neurophysin in the brain and pituitary gland of normal and scrapie-affected sheep. I. Its localisation in the hypothalamus and neurohypophysis, with particular reference to a new hypothalamic neurosecretory pathway to the median eminence. *Neuroscience* **1**, 275–299.
232. Parry, H. B. and Livett, B. G. (1977). Neurophysin in the brain and pituitary gland of normal and scrapie-affected sheep. II. Its occurrence in the cerebellum in dystrophic axon terminals, with lysosome-lipofuscin accumulation. *Neuroscience* **2**, 53–72.
233. Parry, H. B., Tansley, K. and Thomson, L. C. (1955). Electroretinogram during development of hereditary retinal degeneration. *Br. J. Ophth.* **39**, 349–352.
234. Parsons, J. W. (1800). Letter to Sir Joseph Banks of 7th July, quoted by H. B. Carter (1964) pp. 273–274.
235. Pattison, I. H. (1964). The spread of scrapie by contact between affected and healthy sheep, goats or mice. *Vet. Rec.* **76**, 333–336.
236. Pattison, I. H. (1965). Scrapie in the Welsh Mountain breed of sheep and its experimental transmission to goats. *Vet. Rec.* **77**, 1388–1389.
237. Pattison, I. H., Hoare, M. N., Jebbett, J. N. and Watson, W. A. (1972). Spread of scrapie to sheep and goats by oral dosing with foetal membranes from scrapie-affected sheep. *Vet. Rec.* **90**, 465–468.
238. Pattison, I. H., Hoare, M. N., Jebbett, J. N. and Watson, W. A. (1974). Further observations on the production of scrapie in sheep by oral dosing with foetal membranes from scrapie-affected sheep. *Br. Vet. J.* **130**, 65–67.
239. Paulet, J. J. (1775). *Recherches historiques et physiques sur les maladies epizootiques, avec les moyens d'y remédier, dans tous les cas*. 2 Vols, pp. 260–299. Ruault, Paris.
240. Pelham, R. A. (1948). *In: An Historical Geography of England before 1800*. (H. C. Darby, Ed.). Cambridge University Press, Cambridge.
241. Peters, R. A. (1963). *Biochemical lesions and lethal synthesis*. No. 18 of Modern Trends in Physiological Sciences. Pergamon Press, Oxford.
*242. Philipson, D. (1979). Letter to H. B. Parry, dated 31.5.79. (Resö, Sweden.)
243. Power, E. (1941). *Medieval English Wool Trade*. Oxford University Press, Oxford.
244. Pratt, R. T. C. (1967). *Genetics of Neurological Disorders*. Oxford University Press, Oxford.
245. Prély, I. (1854). *Közhasznu Baromorvosi . . .* (a popular account of the care of domestic animals, in Hungarian). G. Müller, Pest.
246. Prestige, M. C. (1974). Axon and cell numbers in the developing nervous system. *Br. Med. Bull.* **30**, 107–111.

247. Raisman, G. (1972). A second look at the parvicellular neurosecretory system *In: Brain-endocrine interaction. Median Eminence: Structure and function* (K. M. Knigge, D. E. Scott and A. Weindt, Eds), pp. 109–118. Karger, Basel.
248. Raynes, F. (1969). Norfolk Horn Sheep. *J. Roy. Agr. Soc. Engl.* **130**, 20–30.
249. Reid, W. L. (1943). Cerebral Oedema. *Aust. N.Z. J. Surg.* **13**, 11–36.
250. Rethelyi, M. and Halász, B. (1970). Origin of the nerve endings in the surface zone of the median eminence of the rat hypothalamus. *Exp. Brain Res.* **11**, 145–158.
251. Richard, P. (1967). *Atlas stéréotaxique du cerveau de brebis 'Préalpes du Sud'.* Paris. Inst. Nat. Rech. Agron.
252. Richthofen, Baron A. K. S. von (1827). *Die Traberkrankheit der Schafe, verglichen mit der sog. Schafraudekrankheit.* Korn: Breslau. Reviewed by J. J. in Bull. Sci. Agric. Econ. (Paris) 1829, Vol. 13, pp. 132–136.
253. Roche-Lubin, M. (1848). Mémoire pratique sur la maladie connue sous les noms de prurigo lombaire, convulsive, trembleuse, tremblante. *Rec. Méd. Vét.* **25**, pp. 698–714.
254. Röll, M. F. (1867). *Lehrbuch der Pathologie und Therapie der Hausthiere.* Wien.
255. Roos, R., Gajdusek, D. C. and Gibbs, C. J. (1973). The clinical characteristics of transmissible Creutzfeldt-Jakob disease. *Brain* **96**, 1–20.
256. Rubios, P. (1436–1517). *Essayo de la Sociedad Vascongoda*, pp. 128–129. Quoted by Klein, 1920, p. 6.
257. Ryder, M. L. (1962–3). Sheep and wool in history. Journal of Bradford Textile Society, pp. 29–43.
*258. Ryder, M. L. (1964). The history of sheep breeds in Britain. *Agric. History Rev.* **12**, 1–12; 65–82.
*259. S. G. L. (Reviewer's initials). Notices of articles by Thaer (1825) and others, in Bulletin des Sciences Agricoles et Economiques (1827), Vol. 7, pp. 217–221.
260. St. Albans Chronicle (1274). Quoted by Trow-Smith (1957).
261. Santos Aran (1944). Quoted by Belda and Trujillano (1979), p. 334.
262. Saper, C. B., Loewy, A. D., Swanson, L. W. and Cowan, W. M. (1976). Direct hypothalamo-autonomic connections. *Brain Res.* **117**, 305–312.
263. Saurat, P. (1941). Contribution a l'étude de la tremblante du mouton. Thesis. Imprimerie Toulousiane.
263a. Schofield, F. W. (1938). A case of scrapie in an imported ewe. Report of the Ontario Veterinary College, pp. 34–35. Ontario Dept. of Agriculture, Toronto.
*264. Schulz, Baron (1800). Observations on sheep, particularly those of Sweden. *Trans. Roy. Dublin Soc.* **1**, 171–207.
*265. Schulze, M. (1967). *Die Anfänge der landwirtschaftlichen Literatur in niedersächsischen Bibliotheken.* Göttingen (dissertation).
266. Schut, J. W. (1950). Hereditary Ataxia: clinical study through six generations. *Arch. Neurol. Psychiat.* **63**, 535–568.
267. Schut, J. W. (1951). Hereditary Ataxia: a survey of certain clinical, pathologic and genetic features with linkage data on five additional hereditary factors. *Am. J. Hum. Genet.* **3**, 93–110.
268. Schut, J. W. and Böök, J. A. (1953). Hereditary Ataxia: difference between progeny of male and female affected members and a definition of certain signs useful in detecting the disease prior to onset of clinical symptoms. *Arch. Neurol. Psychiat.* **70**, 169–179.
269. Schut, J. W. and Haymaker, W. (1951). Hereditary Ataxia: a pathologic study of five cases of common ancestry. *J. Neuropath. Clin. Neurol.* **1**, 183–213.

270. Secretary to Board (1813). *General View of Agriculture of Lincolnshire.* 2nd edn. London.

271. Seitelberger, F. (1962). Eigenartige familiär-hereditäre Krankheit des Zentralnervensystems in einer niederösterreichischen Sippe. *Wien. Klin. Wschr.* **41–42**, pp. 687–691.

*272. Sheep Development Association (1964–1977). Annual reports 1–14.

273. Sheppard, E. (1808). Letter to Sir John Sinclair, Bart., on the subject of his experiments regarding the improvement of the fine-woolled breeds of sheep in this kingdom. Communications to the Board of Agriculture (U.K.), Vol. 6, part 1, pp. 65–73.

274. Short, B. F. and Carter, H. B. (1955). An analysis of the records of the registered Australian merino stud flocks. Bulletin No. 276. CSIRO, Melbourne.

275. Sinclair, J. (ed.) (1792). *Observations on the different breeds of sheep.* Edinburgh.

276. Sjödin, E. (1974). *Får* (transl: sheep) L. T. Förlag, Stockholm. (3rd edn.) Cited by Philipson (1979).

277. Slater, E. (1965). Clinical aspects of genetic mental disorders. *In: Biochemical Aspects of Neurological Disorders* (J. N. Cummings and M. Kremer, Eds), 2nd Series, Ch. 17. Blackwell, Oxford.

278. Smith, R. S. (1928). Medieval Agrarian Society in its Prime: Spain. *Cambridge Economic History of Europe*, Vol. 1, p. 351.

279. Smith, W. T. (1976). Intoxications, poisons and related metabolic disorders. *In: Greenfield's Neuropathology* (W. Blackwood and J. A. N. Corsellis, Eds), 3rd Edn. chap. 4, pp. 148–193. Arnold, London.

280. Snyder, L. M., Necheles, T. F. and Reddy, W. J. (1970). G-6-PD Worcester. A new variant, associated with x-linked optic atrophy. *Am. J. Med.* **49**, 125–132.

281. Spatz, H. (1938). Die systemischen Atrophien. *Arch. Psychiat. Neurol.* **108**, 1–18.

282. Spooner, W. C. (1844). *History, structure, economy and diseases of sheep*, London. (Also 5th edn, 1874.)

283. Stamp, J. T., Brotherston, J. G., Zlotnik, I., Mackay, J. M. K. and Smith, W. (1959). Further studies on scrapie. *J. Comp. Path.* **69**, 268–280.

284. Steele, T. W. (1964). The control and possible eradication of natural scrapie. IV. The practical problems of flockmasters. Rep. Scrapie Seminar, Washington, D.C., U.S.D.A., ARS, 91–53, pp. 316–318.

285. Stevenson, W. (1812). *General View of the Agriculture in the County of Dorset.* London.

286. Stockman, S. (1913). Scrapie: an obscure disease of sheep. *J. Comp. Path. Ther.* **26**, 317–327.

287. Stockman, S. (1926). Contribution to the study of the disease known as Scrapie. *J. Comp. Path. Ther.* **39**, 42–71.

288. Stuart, J. E., Allen, R. K., Schultz, G., Delay, P. D. and Rosenberger, A. C. (1952). An outbreak of scrapie in California sheep. *California Vet.* **6**, 22.

*289. Stumpf, G. (1785). *Versuch einer pragmatischen Geschichte der Schäfereien in Spanien, und der Spanischen in Sachsen.* J. G. Müller, Leipzig. (Translation: An essay on the practical history of sheep in Spain and of the Spanish sheep in Saxony). Transactions of the Royal Dublin Society 1800, Vol. 1, part 1, pp. 1–101).

290. Swanson, L. W. (1977). Immunohistochemical evidence for a neurophysin-containing autonomic pathway arising in the paraventricular nucleus of the hypothalamus. *Brain Res.* **128**, 346–353.

291. Szentágothai, J. (1964). The parvicellular neurosecretory system. *In: The Diencephalon* (W. Bargmann and J. P. Schadé, Eds), *Progress in Brain Research*, Vol. 5, pp. 135–144. Elsevier, Amsterdam.

292. Szentágothai, J. (1970). Glomerular synapses, complex synaptic arrangements, and their operational significance. *In: The Neurosciences: Second Study Program* (F. O. Schmitt, Ed.), pp. 427–442. Rockefeller University Press, New York.

*293. Szent-Iványi, T. (1978). Letter to H. B. Parry, dated 10.8.78. (Dept. of Epizootiology, University of Veterinary Medicine, Budapest.)

294. Takki, K. (1974). Gyrate atrophy of the choroid and retina associated with hyperornithinaemia. *Br. J. Ophthal.* **58**, 3–23.

295. Takki, K. and Simell, O. (1974). Genetic aspects in gyrate atrophy of the choroid and retina with hyperornithinaemia. *Br. J. Ophthal.* **58**, 907–916.

296. Taylor, H. (1934). Scrapie in Sussex. *Vet. Rec.* **14**, 451–452.

297. Tessier, A. H. (1811). *L'Instruction sur les bêtes à laine et particulièrement sur la race des mérinos.* Paris.

298. Thaer, A. D. (1811). *Handbuch für die feinwollige Schaafzucht.* Berlin.

299. Thaer, A. D. (1817). (Sur le vertige des brebis à Frankenfelde) article in *Möglin. Ann. Landwirtsch.* Vol. I, pp. 1–123. Reviewed by S. G. L. in Bull. des Sciences Agric. et Econ. (1827) **7**, pp. 217–219.

300. Thaer, A. D. (1827). Die Traberkrankheit der Schafe. *Moglin. Ann. Landwirtsch.* **19**, 309–318.

301. Thirsk, J. and Imray, J. (1958). *Suffolk Farming in the Nineteenth Century.* Suffolk Records Society, Ipswich, Vol. I.

302. Thompson, B. (1808). A letter to . . . the president of the Newark Agricultural Society on the practicability and importance of introducing the merino breed of sheep . . . Nottingham.

302a. Thorp, F. J., Judd, A. W., Gray, M. L. and Sholl, L. B. (1952). Scrapie in Sheep. *Michigan State Coll. Vet.* **13**, 36–37.

303. Tindall, G. (1977). On Cant, Fashion and Conformity. Encounter, Vol. 48, Ch. 6, pp. 66–71.

304. Trasbot (1890). *Cours de Pathologie Spéciale.* Quoted by Lucam, Béchade and Saurat (1950).

305. Traub, R., Gajdusek, D. C. and Gibbs, C. J. (1977). Transmissible virus dementia; the relation of transmissible spongiform encephalopathy to Creutzfeldt-Jakob disease. *In: Aging and Dementia* (M. Kinsbourne and W. L. Smith, Eds), Ch. 5. Spectrum, New York.

306. Trow-Smith, R. (1957). *A history of British livestock husbandry to 1700.* Routledge and Kegan Paul, London.

307. Trow-Smith, R. (1959). *A history of British livestock husbandry, 1700–1900.* Routledge and Kegan Paul, London.

308. Turner, G. (1795). General view of the agriculture of the county of Gloucester; with observations on the means of its improvement; drawn up for the consideration of the Board of Agriculture and Internal Improvement. *Bath Papers* **7**, 222–252.

309. Tyrrell, D. A. J., Parry, R. P., Crow, T. J., Johnstone, E. and Ferrier, I. N. (1979). Possible virus in schizophrenia and some neurological disorders. *Lancet* **1**, 839–841.

310. Uttenthal, L. O. and Hope, D. B. (1970). The isolation of three neurophysins from porcine posterior pituitary lobes. *Biochem. J.* **116**, 899–909.

311. van Bogaert, L., Dewulf, A. and Pállsson, P. A. (1978). Rida in sheep. Pathological and Clinical Aspects. *Acta Neuropath. (Berlin)* **41**, 201–206.

*311a. Vancouver, C. (1794). *General view of the agriculture in the County of Cambridge.* London.

312. Vancouver, C. (1808). *General view of the agriculture of the County of Devon*, pp. 338–352. London.

313. Vancouver, C. (1813). *General view of the agriculture of Hampshire including the Isle of Wight*, p. 375. London.

314. van den Akker, S., Bool, P. H. and Wensvoort, P. (1968). Scrapie, een chronische aandoening bij het schaap. *Tijdschr. Diergeneesk.* **93**, 898–911.

315. van der Merwe, G. F. (1966). The first occurrence of scrapie in the republic of South Africa. *J. S. Afr. Vet. Med. Ass.* **17**, 415–418.

316. van Klaveren, J. (1960). *Europäische Wirtschaftgeschichte Spaniens Im 16. und 17. Jahrhundert.* p. 200ff. Stuttgart.

317. Wagner, A. R., Goldstein, H. E., Doran, J. E. and Hay, J. R. (1954). Scrapie —a study in Ohio. *J. Am. Vet. Med. Ass.* **124**, 136–140.

*318. Warming, M. (1979). Letter to H. B. Parry, dated 6.7.79. (Veterinaerdirektoratet, Copenhagen, Denmark).

319. Wassmuth, R. (1979). Die Schafrassen. *In: Lehrbuch der Schafzucht* (H. Behrens, H. Doehner, R. Scheelje and R. Wassmuth, Eds), Ch. 3, pp. 124–149. Paul Parey, Hamburg.

319a. Watson, W. A., Terlecki, S., Patterson, D. S. P., Sweasey, D., Hebert, C. N. and Done, J. T. (1972). Experimentally produced progressive retinal degeneration (bright blindness) in sheep. *Br. Vet. J.* **128**, 457–469.

*320. Whitley, R. S. (1977). Letter to H. B. Parry, dated 16.5.77. Grasslands Trial Unit, Stanley, Falkland Islands.

321. Wight, P. A. L. (1960). The histopathology of the spinal cord in scrapie disease of sheep. *J. Comp. Path.* **70**, 70–83.

*322. Willich, A. F. H. (1802). *The Domestic Encyclopaedia, or A Dictionary of Facts and Useful Knowledge*, Vol. III, pp. 495–496. London.

323. Wilson, D. R., Anderson, R. D. and Smith, W. (1950). Studies in scrapie. *J. Comp. Path.* **60**, 267–282.

324. Youatt, W. (1837). *Sheep: their breeds, management and diseases, to which is added the Mountain Shepherd's Manual.* London.

325. Young, A. (1770). *A course of experimental agriculture.* 2 vols. Dodsley, London.

326. Young, A. (1793). Some farming notes in Essex, Kent and Sussex. *Ann. Agric.* **20**, 220–296.

*327. Young, A. (1800). Experiments on the Winter and Summer support of sheep. *Ann. Agric.* **34**, 414–425.

328. Young, A. (1813). *General View of the Agriculture of Oxfordshire.* pp. 298–315. London.

329. Zambrano, D. and de Robertis, E. (1967). Ultrastructure of the hypothalamic neurosecretory system of the dog. *Z. Zellforsch. Mikrosk. Anat.* **81**, 264–282.

330. Zlamál, V. (1859). *Allatgyogyaszat* ... (a book on animal care, for educated people, in Hungarian). Budapest.

331. Zlamál, V. (1877). (*Details of animal illness and its cures, with special regard to academic and veterinary nomenclature.* (In Hungarian), Budapest. (Scrapie discussed on pp. 138–139.)

332. Zlotnik, I. (1957). Vacuolated neurones in sheep affected with scrapie. *Nature (London)* **179**, 737.

333. Zlotnik, I. (1958a). The histopathology of the brain stem of sheep affected with natural scrapie. *J. Comp. Path.* **68**, 148–166.

334. Zlotnik, I. (1958b). The histopathology of the brain stem of sheep affected with experimental scrapie. *J. Comp. Path.* **68**, 428–438.

335. Zlotnik, I. (1963). Experimental transmission of scrapie to golden hamsters (letter) *Lancet* **2**, 1072.

336. Zlotnik, I. (1975). Virus infection and degenerative conditions of the central nervous system. *In: Viral Diseases of the Central Nervous System* (L. S. Illis, Ed.), Ch. 8, pp. 122–144. Baillière Tindall, London.

337. Zlotnik, I. and Katiyar, R. D. (1961). The occurrence of scrapie disease in sheep of the remote Himalayan foothills. *Vet. Rec.* **73**, 543–544.

338. Zlotnik, I. and Rennie, J. C. (1957). The occurrence of vacuolated neurones and vascular lesions in the medullas of apparently healthy sheep. *J. Comp. Path. Therap.* **67**, 30–36.

339. Zlotnik, I. and Rennie, J. C. (1958). A comparative study of the incidence of vacuolated neurones in the medulla from apparently healthy sheep of various breeds. *J. Comp. Path.* **68**, 411–415.

340. Zlotnik, I. and Stamp, J. T. (1965). Scrapie in a Dorset Down ram. *Vet. Rec.* **77**, 1178.

341. Zlotnik, I. and Stamp, J. T. (1966). The transmission of scrapie from a Dorset Down ram to a Cheviot sheep. *Vet. Rec.* **78**, 222.

Index

187